Introductory Circuits

Introductory Circuits

Robert Spence

Imperial College London, UK

A John Wiley and Sons, Ltd., Publication

This edition first published 2008
© 2008, John Wiley & Sons, Ltd

Registered office
John Wiley & Sons Ltd, The Atrium, Southern Gate, Chichester, West Sussex, PO19 8SQ, United
Kingdom

For details of our global editorial offices, for customer services and for information about how to
apply for permission to reuse the copyright material in this book please see our website at
www.wiley.com.

The right of the author to be identified as the author of this work has been asserted in accordance
with the Copyright, Designs and Patents Act 1988.

Reprinted with corrections August 2009

Library of Congress Cataloging-in-Publication Data

Spence, Robert, 1933–
  Introductory circuits / by Robert Spence.
      p. cm.
  Includes index.
  ISBN 978-0-470-77971-2 (pbk.)
 1. Electronic circuits.   2. Electric circuits.   I. Title.
  TK7867.S639 2008
  621.3815–dc22
                              2008012272

British Library Catalogue in Publication Data

A catalogue record for this book is available from the British Library

ISBN 978-0470-77971-2

Typeset in 10/12 pt Times by Aptara Inc., New Dehli, India

To the 4000 students who have taken my first-year circuits courses, and thereby taught me a lot about circuits. Forgive me if I don't mention you by name.

Contents

About the Author

Bob Spence has been teaching courses on electronic circuits for about 45 years, first to mainstream Electrical Engineering students at Imperial College London and lately to Biomedical Engineering students at the same institution. His research in electrical engineering and human–computer interaction has been recognized by the award of three doctorates and by election to Fellowship of the Royal Academy of Engineering. Bob is Professor Emeritus of Information Engineering, and a Senior Research Investigator, at Imperial College London.

Preface

Some years ago I declared to my colleagues that I would never, *ever,* write a first year textbook. So what happened?

What happened was that I could find no textbooks appropriate to the 20-lecture course on circuits that I was teaching. Why? There were many reasons. Most texts were far too large, heavy and expensive: 1000 pages? 2 kg? £70? No thanks. Many were devoted exclusively to *linear* circuit theory, ignoring the many useful circuits that my students would later encounter. Some decided to teach some of the mathematics required: but my mathematics colleagues do that far better than I. Others decided to treat the underlying physics: I prefer to make a well-defined start with the relations imposed on currents and voltages by components and connections (otherwise, where do you stop? Back emf? Schrödinger's equation?). Some authors are exhaustive (and exhausting): is it really necessary to teach mesh analysis (which no one uses) to students who are not going to be dedicated circuit theorists in later life?

So I prepared and modified my own 'handout' notes which, eventually, started looking like the book I had been searching for. What you have in your hand is essentially those notes, supplemented by quite a number of worked examples which students always find useful and always request.

At the end of each chapter I have included useful problems with a selection of answers. The remaining solutions can be found on a companion website at www.wiley.com/go/spence_circuit.

I used to teach mainstream electrical engineering undergraduates and I now teach non-EE students, specifically those pursuing a course in Biomedical Engineering. It may well be, therefore, that the book is particularly suited to 'non-EE' first-year students, though I suspect that it may still be useful as a supporting text for mainstream EE undergraduates.

Robert Spence

The Design Process

The principal reason for learning how to analyse the behaviour of a circuit is that we shall eventually want to design one or try to understand one that has been designed. So to provide a context for the entire book we look briefly at the design process to see where the material of this book fits in and where it doesn't.

Figure 1.1 provides an overview of the circuit design process. Someone (**A**), somewhere – and it may be you, the circuit designer – needs a circuit designed. You must therefore say what performance you want from the circuit: in other words, you provide specifications (**B**) for its behaviour. The performance may, for example, be the extent to which the voltage captured from an aerial must be amplified to operate a loudspeaker. You, the circuit designer (**C**), must propose an idea (**D**) for a circuit that will exhibit the required performance. This is the hard part! You may achieve this by recalling a circuit designed earlier, and try to modify it; or you may consult a book to find what might be a suitable circuit; or you may ask a fellow engineer for ideas; or you may simply draw upon your experience of circuit design and create a circuit from scratch.

The idea will usually be sketched as a circuit diagram on a piece of paper so that it can be reviewed. But what then? There are three routes that may be followed.

One approach is to build the circuit (**E**) and then measure its performance (**F**). The measured performance (**G**) can then be compared (**H**) with the customer's specification. If the two agree then circuit design might stop at that point. But it would be remarkable if the first idea for a design were satisfactory. In that case you, the designer (**C**) must decide how to modify (**J**) the circuit (or discard it and start again). The circuit would be rebuilt, measured again, and its performance checked against specifications. Again, the circuit may not work exactly as required, so the loop **C–D/J–E–F–G–H–C** is traversed once more. Indeed, it would not be unusual for this loop to be traversed many times in order to ensure a well-designed circuit.

Introductory Circuits   Robert Spence
© 2008 John Wiley & Sons, Ltd

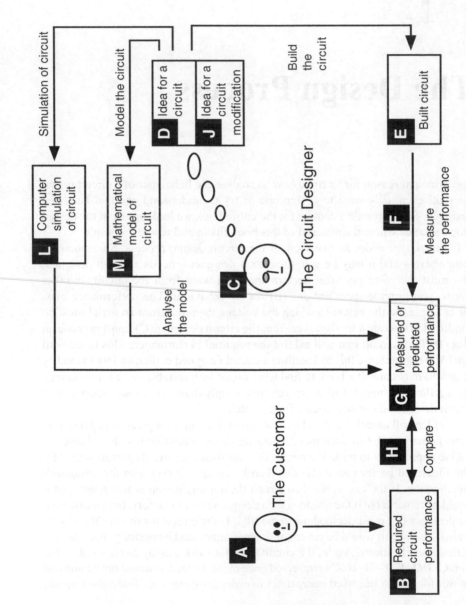

Figure 1.1   An overview of the process of circuit design

The approach we have just described is totally inappropriate if you are designing an integrated circuit: the cost of manufacturing such a circuit is so high and, moreover, would be multiplied many times if the first few attempts at design were not satisfactory. In this case you would use one of the many available software packages (**L**) to simulate the proposed circuit's performance and compare that prediction (**G**) with the specifications (**H**). Again, the design loop (here, **C–D/J–L–G–H–C**) will probably be traversed a number of times until the simulated circuit performance (**G**) satisfies the specifications (**B**). At that point, with some degree of confidence, a decision may be made to manufacture the circuit.

A third approach is for the circuit designer to write down a mathematical model (**M**) – normally a number of equations – describing the proposed circuit, and then to solve those equations in order to find the proposed circuit's performance (**G**). Again, because the initially proposed circuit may not quite meet the specifications an iterative process may be required.

In what way does this book prepare one to undertake design, and in what way does it not? Its principal value relates partly to the transition from **C** to **D**: in other words the process by which you, the designer, propose what you think might be a suitable circuit. The material of this book, and especially the solution of problems, should provide some experience relevant to this initial stage of the design process. Another principal value of the book is that it shows how a circuit (**D**) can be modelled (**M**) and then analysed to find its performance (**G**), thereby enabling a wide variety of circuit performances to be investigated.

This book does *not* address the task of building a circuit (**E**) and measuring (**F**) its behaviour: such skills are usually acquired in a laboratory course. Equally, it does not address the task of using software (**L**) to simulate circuit performance: again, special classes are often organized to introduce students to this task.

2

Electronic Circuits

It is stating the obvious to say that electronic circuits are essential components of computation, instrumentation, communication and many other vital systems. It is therefore necessary for us to be able to determine how a given circuit will perform and, of course, to be able to invent new circuits to perform new functions. The first question is how we *describe* these circuits.

2.1 Voltage and Current

A circuit comprises a number of interconnected parts, each of which imposes its own unique relationship between two electrical quantities – *voltage* and *current*. It is these relationships, and their variety, which makes it possible for the creative circuit designer to design many different useful circuits. Once we know how the parts are connected to form a circuit we can then describe that circuit (essentially a new 'component') in terms of the relationships that *it* now imposes upon its voltages and currents.

What can we usefully say about these variables called 'voltage' and 'current'? It is sometimes helpful to draw an analogy between electrical and hydraulic systems: electrical current is similar to the flow of water, and voltage is similar to water pressure. Given a pipe in which water can flow (Figure 2.1), the rate of flow will be governed by the difference in water pressure between its two ends. The dimensions of the pipe will determine the relation between the water flow rate and the difference in pressure between the two ends: a thinner tube will resist the flow of water, whereas a wider one will allow the water to flow faster. In the same way (Figure 2.2) an electronic component will determine the relation between the voltage across it and the current through it.

Introductory Circuits   Robert Spence
© 2008 John Wiley & Sons, Ltd

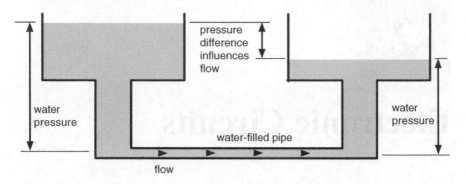

Figure 2.1   Illustrating the water pressure/flow analogy of voltage and current

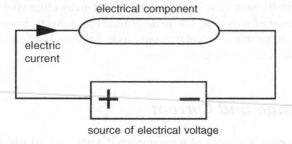

Figure 2.2   Application of a voltage across a component causes a current to flow

It can also be helpful to ask exactly what electric current is. Briefly, and without going into the physics involved,[1] electric current is the movement, in a piece of material, of those electrons within the material that are available for movement. In materials such as copper, silver and most metals there are many free electrons, so that a relatively small voltage will lead to a substantial current. Other materials such as plastic and glass are known as insulators and have few free electrons, so that only a miniscule current flows, even when a very high voltage is applied. Conventionally, we use the name 'current' to refer to movement of electrons in the opposite direction (Figure 2.3).

_____

[1] The omission of any mention of the physics underlying electrical behaviour is intentional. There are various levels at which electrical behaviour can be described. The amplifier that connects your CD to the loudspeaker can be described, at one extreme, as a 'black box' introducing a specific voltage amplification, or by the various voltages and currents in its internal components, or by the physics underlying the different components in the circuit. A line has to be drawn somewhere according to the interest of the person dealing with the circuit. Our view in this book concerns analysis leading to design, and it for this reason that the line is drawn above the physics.

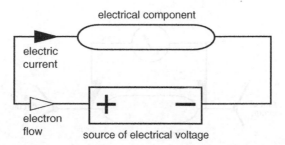

Figure 2.3 Negative free electrons are attracted to the positive potential of the source. However, in all discussions of electrical circuits the conventional current is regarded as flowing in the opposite direction

What is voltage? The obvious answer is that it is that which encourages the free electrons to flow. In many cases a voltage is produced by chemical means, as in conventional batteries: in others – as we shall see in Chapter 12 – it is produced by an electronic circuit.

The difference between current and voltage can also be related to the way in which they are measured. Current, measured in *amperes* (or simply amps), is usually measured by an ammeter (Figure 2.4) through which the current to be measured actually flows. Usually one wants to measure current without disturbing the circuit within which it is connected – that is, without impeding the current in any way. For this reason, ammeters are usually designed to have what is called *low impedance*. Voltage, measured in *volts*, is usually measured (Figure 2.5) by a voltmeter whose two terminals are connected to the points in the circuit whose voltage difference is to be measured. As with the ammeter, to avoid the connection of a voltmeter disturbing a circuit, voltmeters are designed to have a high impedance – that is, they impede, as far as possible, the diversion of any current through the voltmeter.

With the development of technology, present-day ammeters and voltmeters often look a little different (Figure 2.6) from the 'dial and pointer' instrument. There is also a frequent need to observe voltages and currents that are varying with time, and it is here that the oscilloscope (Figure 2.7) finds application.

Figure 2.4 An ammeter indicates the value of the current I flowing through it

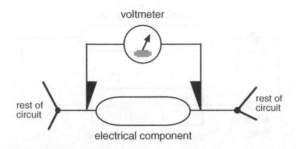

Figure 2.5 A voltmeter indicates the value of the voltage between two points in a circuit

Figure 2.6 A modern 'multimeter' capable of providing a digital indication of voltage and other quantities

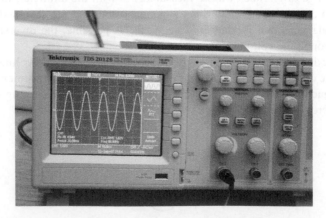

Figure 2.7 An oscilloscope allows time-varying voltages to be observed

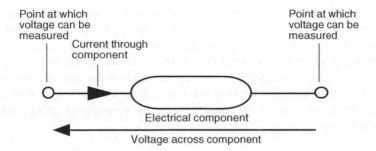

Figure 2.8   Conventional indication of current and voltage associated with an electrical component

We often need to discuss voltages and currents, and give them names. Currents are denoted by arrows superimposed on the wire carrying that current (Figure 2.8), while voltage is indicated by an arrow stretching between the two terminals at which voltage is measured (Figure 2.8). Further details about the meaning of the arrows in Figure 2.8 will be provided when appropriate.

2.2 Power

There is another important electrical variable, called power. *Power* is the rate at which energy is supplied to something. If we consider the 'black box' (called 'black' because we don't know what's inside it!) shown in Figure 2.9, where the voltage between the box's terminals is V and the current entering one terminal and leaving the other is I, then physics tells us that energy is *supplied* to the box at a rate equal to the product of V and I, this rate having the unit of watts. The bulb from the headlight of a car may, for example, be rated at 48 watts, so if the car's voltage supply from its battery is 12 volts we know that the current flowing in the bulb is 48 watts divided by 12 volts, which is 4 amps.

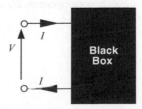

Figure 2.9   Relevant to a definition of the supply of energy to a two-terminal black box

For some components such as resistors, diodes and Zener diodes, V is always positive if I is positive, so energy can only be supplied *to* these components – we call them *passive components* because they cannot supply energy themselves. To supply energy a component would have to be *active*, a property we discuss later. What happens to the energy supplied to resistors and similar passive components? It is dissipated as heat, and because heat can, if there is too much of it in a small area, destroy a component, it is normal to have a maximum *power rating* associated with a component.

2.3   Circuit Diagrams

The way in which components are connected together to form a circuit is usually described by a circuit diagram, an example of which is shown in Figure 2.10. The lines between the components indicate the wiring – the 'connecting together' – of the components. The implication is that the voltage is the same at all points along the wire. In other words, in Figure 2.10, similarly labelled parts of components and wires (e.g., A, A, A.) are all at the same voltage. The black 'blobs' in the circuit diagram of Figure 2.10 confirm the connection of wires at the points indicated. Occasionally we use white blobs to indicate the points at which one circuit with a well-defined function is connected – or available to be connected – to another.

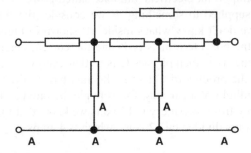

Figure 2.10   A circuit diagram shows how components are connected together. Every point (e.g., A) along a wire has the same voltage

Overview: DC Circuits

If we're going to design electrical circuits we must be able to predict how they will behave: in other words, what their performance will be. That is why we first study circuit analysis though we shall sometimes apply it to circuit design.

Circuit analysis is sometimes carried out by computers and sometimes by human beings. The analysis of a large circuit such as an integrated circuit on a chip must be carried out by computer if the result is needed without delay and must be free from the sort of errors that human beings often make. But the human designer of circuits must be able to analyse simple circuits, often 'on-the-fly' as he or she designs them, often informally and without recourse to a computer, simply because an ability to design relies heavily upon an ability to analyse.

We first consider what are called 'DC circuits' in which all currents and voltages have constant values (DC = direct current). We start with *linear resistive circuits* which contain two types of component, *sources* and *resistors*.

The behaviour of a DC circuit is governed by three sets of equations:

- Kirchhoff's current law

- Kirchhoff's voltage law

- Component current~voltage relations: including Ohm's law and the constant values of current and voltage sources.

These sets of relations, which are *linear* and therefore straightforward to solve, will be illustrated by the first circuit examples we meet. However, after one or two informal attempts at the analysis of simple circuits we realise that there is a need for a systematic method of analysis that can be applied straightforwardly by either a human being or a computer to a circuit of any size. One such method is called the

Introductory Circuits   Robert Spence
© 2008 John Wiley & Sons, Ltd

nodal analysis method. Linearity also means that the *principle of superposition* can sometimes simplify analysis, as can *Thevenin's theorem.* The latter allows us to represent a very complicated circuit by a simple combination of a source and a resistor, thereby making later circuit analysis far simpler.

We then extend our study of DC circuits to include controlled sources, where a voltage at one location in a circuit controls the current flowing in another part of the circuit. The reason we study these 'dependent sources' is that we need them if we are going to be able to explain what happens in an amplifier or switch. In fact, you can regard transistors as attempts to realize controlled sources.

The advantages conferred by linearity disappear when a *nonlinear component* such as a *diode* is part of a circuit. Nevertheless, we can cope with this problem by using what is called the *load-line* approach to circuit analysis, an approach which employs drawing rather than equation solution.

You may ask whether it is worth studying DC circuits in which all voltages are constant and therefore never vary: they sound a little boring. The answer is 'yes', for one very good reason among many. We shall find later that the analysis of much more interesting circuits such as amplifiers *employs precisely the same approach*, so that a knowledge of DC circuit analysis is a good investment.

3

Circuit Laws and Equivalences

To say that

'a circuit is an interconnection of components'

seems rather pointless, but it is not. It is useful because it emphasises the fact that interconnections and components are independent of each other, and described by two different sets of equations. We consider components first in order to establish the relations they impose upon their voltages and currents.

3.1 Components

Resistance

The linear resistor has two terminals and is so called in view of the linear relation between the current that flows through it and the voltage across it. Measurement of a resistor's current and voltage might provide the graph shown in Figure 3.1, depicting a linear relation between the two:

$$V = RI \tag{3.1}$$

known as *Ohm's law*. The constant R is called the *resistance* of the resistor. In the example shown the value of R is 2 ohms (2 Ω).

The straight line will not go on forever, since at high voltages and/or currents the resistor will be destroyed. Nevertheless, because we design circuits to avoid

Introductory Circuits   Robert Spence
© 2008 John Wiley & Sons, Ltd

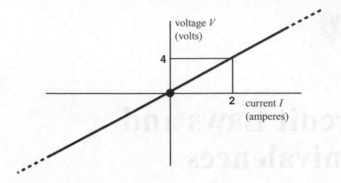

Figure 3.1   The result of measuring the current through, and the voltage across, a resistor

such destruction, we assume a linear relation and reap the benefit of the resulting linear equations that may describe the circuit containing the resistor.

If the circuit designer is drawing a circuit diagram, the resistor whose relation between current and voltage is shown in Figure 3.1 will be represented by the symbol and associated value of resistance shown in Figure 3.2.

Conductance

We can rearrange Ohm's law to get the equivalent relation:

$$I = GV \qquad (3.2)$$

where G is known as the **c**onductance of the resistor, has the units of siemens and, as seen from Equation (3.1), is the reciprocal of the resistance R. For the same resistor whose voltage~current relation is depicted in Figure 3.1 the conductance is 0.5 siemens (0.5 S) and the symbolic representation of the resistor (Figure 3.3) is the same as in Figure 3.2. In Figure 3.3 and henceforth in this book we omit the

Figure 3.2   Symbolic representation of the resistor whose voltage~current relation is shown in Figure 3.1

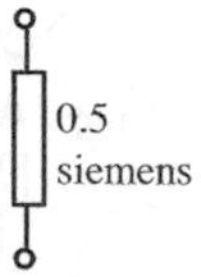

0.5
siemens

Figure 3.3   Alternative symbolic representation of the resistor whose voltage~current relation is shown in Figure 3.1

reference voltage and current, it being understood that they are related as shown in Figure 3.2 (with the reference current entering at the node where the voltage is greater). The reason for considering two equivalent descriptions of a resistor – its resistance and conductance – is that in circuit analysis one may be more convenient to handle than the other.

Reference directions

Let's return to the voltage and current associated with the resistor (Figure 3.2). When analysing a circuit we will want to know, for example, not only the value of the current flowing through a resistor, *but its direction as well*. And the problem is that before we analyse a circuit we usually have no idea whatsoever of the directions in which current will flow in the various components and which terminal of a component will be at a higher voltage than the other. So we acknowledge this fact and simply assign *reference directions* to current and voltage. Take, for example, the resistor shown in Figure 3.4, which is connected to other components to form a circuit. We do not know in which direction its current will flow so we *arbitrarily* choose a reference direction indicated by an arrow, and call the current in that direction I. If it turns out that the current does flow in that direction then I will have a positive value, for example 2 A. If it turns out, however, that the current actually flows in the opposite direction then I will have a negative value, and we

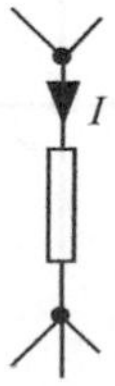

Figure 3.4   The reference direction for current can be arbitrary, and does not necessarily indicate the actual direction of flow

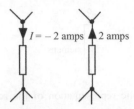

Figure 3.5 If the value of *I* in Figure 3.4 is negative, that can be represented in either of the two ways shown here

could indicate that in one of the two ways shown in Figure 3.5. A negative value for *I* doesn't mean that we have made a mistake: the arrow was only chosen to indicate a *reference* direction, not the *actual* direction of the current flow.

The same comment applies to voltages. In general we do not know, before analysing even the simplest circuit, which terminal of a component will be at a higher voltage than the other. We therefore arbitrarily assign reference directions (using the arrowhead to denote the higher voltage) and see afterwards if the voltage is positive or negative with respect to that reference direction.

We denote the unit of voltage by V. However, to avoid possible confusion with *V* denoting a variable, the term 'volts' is often employed in place of V.

Sources

An *ideal voltage source* is an electrical component characterized by the fact that the voltage across it is constant, and independent of the value or direction of the current flowing through it. Thus, if we could make measurements of the current and voltage associated with an ideal voltage source we would get a result like the example shown in Figure 3.6. It is important to realise that the current can flow in either of the two possible directions. Consideration of a car battery

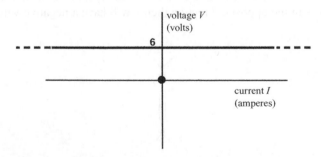

Figure 3.6 The result of measuring the current through, and the voltage across, an ideal voltage source

Figure 3.7 Representation of an ideal voltage source of 6 V

(which approximates to an ideal voltage source) will show that this property is not surprising: when starting the car a current flows 'out of' the battery to turn the starter motor, but once the motor is running, current flows in the opposite direction to charge the battery.

Conventionally, an ideal voltage source is represented by the symbol shown in Figure 3.7. The larger of the two horizontal lines denotes the terminal which is at the higher voltage. Corresponding to the example of Figure 3.6 a label – here '6 V' – indicates the extent of the voltage difference.

A special case of an ideal voltage source is a *short-circuit* (Figure 3.8) whose voltage is always zero, again irrespective of the current flowing through it. In fact, we often use a short-circuit – in the form of a connecting wire - to ensure that two points in a circuit have the same voltage. When looking at a circuit diagram, therefore, one should be aware that the voltage has the same value at any point along a line joining components.

The other ideal source we shall be concerned with is the *ideal current source*. This component is characterized by the fact that the current through it is constant, irrespective of the value and direction of the voltage across it. Thus, if we could make measurements of the current and voltage associated with an ideal current source we would get a result like the example shown in Figure 3.9. Conventionally, an ideal current source is represented by the symbol shown in Figure 3.10. Corresponding to the example of Figure 3.9 a label – here '2 A' – indicates the value of the constant current, and the arrow forming part of the symbol indicates the reference direction for current. Just as a short-circuit is a special case of an ideal voltage source, ensuring that no voltage exists between two points in a circuit, an *open-circuit* is a special case of an ideal current source in which no current flows (Figure 3.11).

A car battery was quoted as an approximation to an ideal voltage source. An ideal current source is frequently approximated in integrated circuits by means of a transistor.

Figure 3.8 A short-circuit. The voltage V between the terminals is zero whatever the value of the current through it

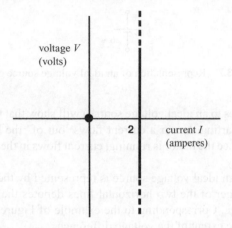

voltage *V*
(volts)

2

current *I*
(amperes)

Figure 3.9 The result of measuring the current through, and the voltage across, an ideal current source

2 A

Figure 3.10 Representation of an ideal current source of 2 A

O X

O Y

Figure 3.11 There is an open-circuit between terminals X and Y, through which no current can flow

Power

In Chapter 2 we introduced the concept of power, and stated that if a black box has a voltage and current as shown in Figure 3.12, where *V* and *I* are both positive, then the power supplied to the black box is *VI* watts. If the black box happens to be a single resistor, then it is clear (see Figure 3.1) that *V* and *I* always have the

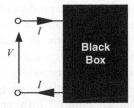

Figure 3.12 The power supplied to a black box is the product of V and I provided the current I enters at the terminal with the highest voltage (i.e., the positive reference for V)

same sign and therefore a resistor can only absorb power: it cannot supply it. For that reason it is called a *passive component*. What happens to the energy supplied to a resistor? It is dissipated as heat; in other words the temperature of the resistor rises. Too much heat will destroy a resistor, which is why a given resistor will carry a maximum power rating that should not be exceeded. Because we know the relation (Equation 3.1) between V and I for a resistor we can write that the power P supplied to a resistor is given by

$$P = VI = RII = I^2R$$

Alternatively, the power supplied

$$P = VI = VV/R = V^2/R$$

Since a voltage source is a two-terminal component we can say something about its ability to supply or receive power (Figure 3.13a). As the characteristic of the

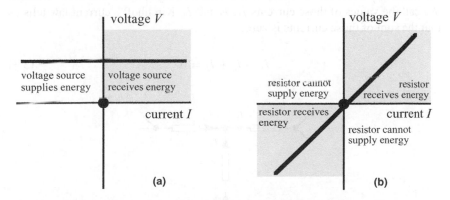

Figure 3.13 A voltage source can supply energy because the product of V and I can be negative. With a resistor the product of V and I is always positive, so it can only receive energy

voltage source shows, it is possible for the product of V and I to be negative and energy to be supplied *by* the source: in view of this potential it is known as an *active component*. It is also possible for the product of V and I to be positive, with energy being supplied *to* the source, the component then acting in a passive manner. This is consistent with our understanding of the familiar car battery as an approximation to a voltage source: it delivers energy when starting the car and receives it when it is being charged. Figure 3.13(b) provides a reminder of the fact that, for a resistor, the product VI must always be positive, and energy can only be absorbed.

3.2 Interconnections

We stated earlier that 'a circuit is an interconnection of components'. Having studied some simple components and the relations they impose upon their currents and voltages we now examine the consequences of connecting them together. There are two consequences, and both are described by laws first presented by the mathematician Kirchhoff.

Kirchhoff's current law

Let's imagine (Figure 3.14) that three resistors are connected together within a circuit so that they share a common terminal, often called a *node*. We do not know in which direction the component currents will flow, so we arbitrarily choose all the reference directions to describe currents flowing into the common terminal. We call the values of those currents I_1, I_2 and I_3. Kirchhoff's current law tells us that the sum of those currents is zero:

$$I_1 + I_2 + I_3 = 0 \tag{3.3}$$

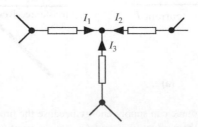

Figure 3.14   Three resistors are connected to the same terminal

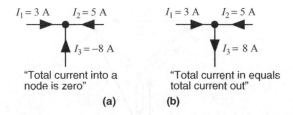

"Total current into a   "Total current in equals
node is zero"   total current out"

(a)   **(b)**

Figure 3.15   Alternative expressions of Kirchhoff's current law

In other words, the total current that flows into a node is equal to the total current flowing out. For example, if I_1 and I_2 are positive, so that these currents flow into the common node, then I_3 will be negative, indicating that current flows away from the common node through the lower resistor. If you find analogies helpful, then imagine the resistors to be replaced by pipes carrying water, such that the flow into their junction will equal the flow out of the junction.

Kirchhoff's current law – usually abbreviated to KCL – applies not only to three interconnected components, but to all the currents entering a node from any number of components, and can be expressed generally as

$$\sum_{\text{node}} I = 0 \tag{3.4}$$

In our illustrative example all the reference currents flowed into the node. KCL applies equally if all the reference currents flow away from the common node. When analysing a circuit you may find it convenient either to state that the sum of the currents into a node is zero (Figure 3.15a) or, equivalently, to say that the total current flowing in is equal to the total current flowing out (Figure 3.15b). Note that KCL does not make any reference to the components carrying the various currents: it is concerned *only* with their interconnection. Indeed, the rectangular symbols shown in Figure 3.14 need not represent resistors: they could equally well represent lumps of cheese.

Kirchhoff's voltage law

Figure 3.16 shows part of a circuit. We shall assume that we have no knowledge of the currents and voltages in the circuit so we arbitrarily name the various component voltages as V_A, V_B and V_Y as shown, and we have used V_X to denote the voltage between two nodes that are not directly connected by a component. Kirchhoff's voltage law – usually abbreviated to KVL – tells us that if we trace a connected sequence of reference voltages that forms a loop (for example, V_A,

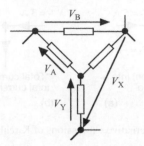

Figure 3.16   A closed loop is formed by the voltages V_A, V_B, V_X and V_Y

V_B, V_X and V_Y) then the sum of all those voltages, taking reference directions into account, is equal to zero. Thus, for the example of Figure 3.16,

$$V_A + V_B + V_X + V_Y = 0 \qquad (3.5)$$

To emphasize the need to be careful about signs we examine another circuit (Figure 3.17). Paying attention to voltage reference directions, the application of KVL to this circuit will yield

$$V_C - V_D + V_R - V_S = 0 \qquad (3.6)$$

It is very important to note that the voltages involved in an expression of KVL need not be voltages measured directly across components: see, for example, V_X in Figure 3.16 and V_S in Figure 3.17. It is also important to realise that, like KCL, KVL makes no reference to components; *it is concerned only with interconnections*. In the examples of Figures 3.16 and 3.17 the components are resistors; they could equally well be a mixture of resistors and sources.

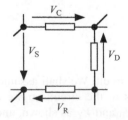

Figure 3.17   Four voltages forming a closed loop within a circuit

Table 3.1   Summary of the relations describing DC circuits

FEATURES OF A CIRCUIT	SYMBOLIC REPRESENTATION	RELATION
Components		Ohm's Law constant V constant I
Connection at a node		Kirchhoff's Current Law (KCL)
Connection to form a loop		Kirchhoff's Voltage Law (KVL)

Just as Kirchhoff's current law can be expressed generally, so can KVL, as

$$\sum_{\text{loop}} V = 0 \qquad (3.7)$$

where we use the term 'loop' to denote any closed connection of voltages.

Summary

To emphasize the fact that a DC circuit containing only sources and linear resistors is described by only three sets of relations we provide Table 3.1. There is a second, and very important reason for providing this table. When we later come to consider AC circuit behaviour (Chapters 9–11) and change behaviour (Chapters 12, 13) we shall find that they are described by the same type of relations as for DC circuits, thereby immensely simplifying the task of understanding and analysis.

3.3 Equivalence

Very often we find, in a circuit, two resistors connected in series (Figure 3.18) and it helps enormously if we can represent that connection as a single resistor. A similar situation arises when two resistors are connected in parallel (Figure 3.19).

Figure 3.18   The series connection of two resistors. Note that there is nothing else connected to point X

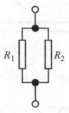

Figure 3.19   The parallel connection of two resistors

To find the value of a single resistor which has the same electrical characteristic as two resistors in series we assume (Figure 3.20) that the current through R_1 is I. All of this current must also flow through R_2 (there is nowhere else for it to go – a simple application of KCL at node X). From Ohm's law we know that the voltages across R_1 and R_2 are $R_1 I$ and $R_2 I$, respectively. Using KVL we can also write

$$V = R_1 I + R_2 I = (R_1 + R_2)I \tag{3.8}$$

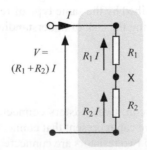

Figure 3.20   Derivation of the equivalent resistance of two resistors connected in series

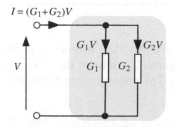

Figure 3.21 Derivation of the equivalent conductance of two resistors connected in parallel

We therefore have a 'new component' indicated by the shaded area in Figure 3.20 whose relation between voltage V and current I (Equation 3.8) is identical in form to Ohm's law and therefore describes a resistor having a value of $R_1 + R_2$, the sum of the two original resistances. Therefore, wherever two resistors are connected in series they can be replaced by a single equivalent resistor. Note, however, that the equivalence depends upon the same current flowing through R_2 as flows through R_1. As a consequence, if anything were to be connected to the node X in Figure 3.20 and draw current from it, the replacement of the two series-connected resistors by a single equivalent resistor would not be valid.

In the same way it can be shown that two resistors in parallel can be replaced by a single equivalent resistor (Figure 3.21). In terms of conductance, the equivalent conductance is simply the sum of the two separate conductances. Thus,

$$G_{eq} = G_1 + G_2 \qquad (3.9)$$

If one prefers to work in terms of resistance the expression for the equivalent resistance is

$$R_{eq} = 1/G_{eq} = R_1 R_2/(R_1 + R_2) \qquad (3.10)$$

where $R_1 = 1/G_1$ and $R_2 = 1/G_2$.

3.4 Simple Circuit Analysis

To illustrate the application of the three sets of relations (Table 3.1) describing DC circuits we examine four simple circuits. They are not trivially simple, because they illustrate concepts that will be needed when predicting the behaviour of more complex circuits.

Example 3.1

Here (Figure 3.22) we have two ideal sources connected together, and we have to find the value of the current I through the voltage source and the voltage V across the current source. The reference directions of V and I have been chosen arbitrarily.

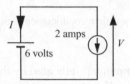

Figure 3.22   The circuit analysed in Example 3.1

The current I must take on the value $-2\,\text{A}$ because the current source is a statement that the current flowing in it, and therefore through the voltage source, is fixed at 2 A irrespective of the voltage across the current source. The minus sign occurs because we have arbitrarily chosen the reference direction for I as shown in the figure. Similarly, the voltage V across the current source is defined by the voltage source to be 6 V because the voltage source is connected directly across the current source. Recall that in this ideal circuit there is no voltage drop across the wires joining the two sources.

Example 3.2

In Figure 3.23 we have two ideal current sources connected to a resistor and we want to find the voltage V across the resistor. We first invoke KCL at the node

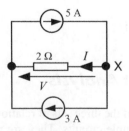

Figure 3.23   Pertinent to Example 3.2

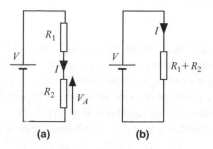

Figure 3.24   Pertinent to Example 3.3

X to find the current I $(= 5 - 3 = 2\,\text{A})$ flowing in the resistor. Now applying Ohm's law we find the voltage $V = -2\,\text{A} \times 2\Omega = -4\,\text{V}$. Again, the minus sign arises from our choice of reference direction for V. The minus sign of V means that the voltage at point X is higher than the voltage at the other terminal of the resistor.

Example 3.3

The simple circuit of Figure 3.24(a) will often be encountered within the circuits we examine. Let us suppose that we have to calculate the value of the voltage V_A across the resistor R_2. To do that using Ohm's law we need to know the current through R_2. That current I is also the current through R_1. To find I we can replace the series connection of R_1 and R_2 by a single resistor of value $R_1 + R_2$ (Figure 3.24b). The value of I now follows from Ohm's law:

$$I = V/(R_1 + R_2) \tag{3.11}$$

Knowing I, we can now go back to Figure 3.24(a) and use Ohm's law to find V_A:

$$V_A = R_2 I = V[R_2/(R_1 + R_2)] \tag{3.12}$$

This relation is worth remembering because the circuit of Figure 3.24(a), often referred to as a 'voltage divider', is widely used.

Example 3.4

What can be called the 'dual' of the circuit in Example 3.3 is shown in Figure 3.25(a): a current source feeding into two resistors connected in parallel. To find the currents I_1 and I_2 we can proceed as follows. First we replace the two parallel resistors by a single equivalent resistor as shown in Figure 3.25(b). According to Equation (3.10) the value of that resistor is $6 \times 3/(6+3) = 2\Omega$. A current of 6 A flowing through that resistance will, by Ohm's law, create a voltage V given by

$$V = 6 \times 2 = 12V$$

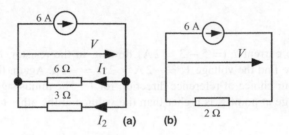

Figure 3.25   Pertinent to Example 3.4

Now that we know the voltage V we can return to the original circuit of Figure 3.25(a) and use Ohm's law to find I_1 ($=12\,V/6\Omega = 2\,A$) and I_2 ($=12\,V/3\Omega = 4\,A$). As a check (always useful to carry out!) we note that the sum of these currents is 6 A, thereby confirming that KCL is obeyed at the junction of the two resistors and the current source.

The circuit of Figure 3.25(a) is useful to clear a misconception that frequently occurs. It is often erroneously thought that 'the current takes the path of least resistance', whereas it simply divides in keeping with the three circuit laws of Table 3.1. There is no way that an 'intelligent' current looks ahead and then decides what route to take.

3.5 Problems

Simple circuit analysis

Problem 3.1
Figure P3.1 contains 20 components and simple circuits. For each one find the unknown current and/or voltage as indicated with a question mark.

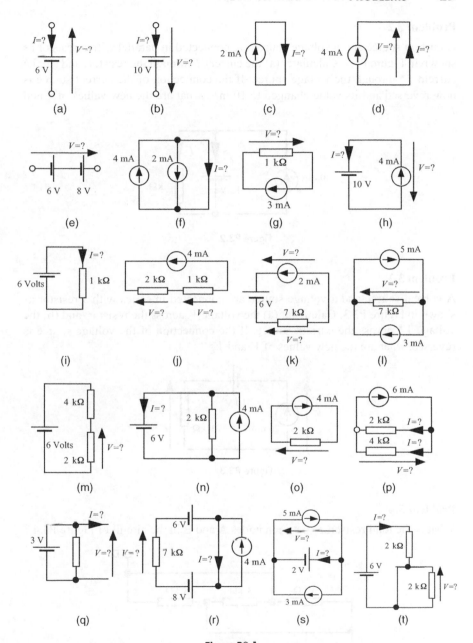

Figure P3.1

Problem 3.2

A current source and a voltage source are connected in parallel with a resistor as shown in Figure P3.2. Calculate: (a) the current I through the resistor; and (b) the current I^* through the voltage source. If the connection of the current source is now reversed and its value changed to 10 mA, what are the new values of I and I^*?

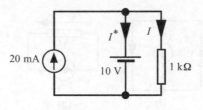

Figure P3.2

Problem 3.3

A current source and a voltage source are connected in series with a resistor as shown in Figure P3.3. Calculate: (a) the voltage V across the resistor: and (b) the voltage V^* across the current source. If the connection of the voltage source is reversed, what are the new values of V and V^*?

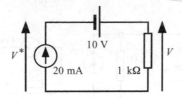

Figure P3.3

Problem 3.4

What is the total resistance between nodes A and B in the circuit of Figure P3.4 ?

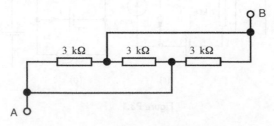

Figure P3.4

Problem 3.5

By using appropriate equivalence relations find the value of the currents I_A and I_B in the circuit of Figure P3.5.

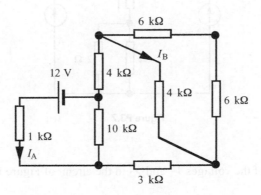

Figure P3.5

Problem 3.6

For the circuit shown in Figure P3.6 find the values of the voltages V_1 and V_2.

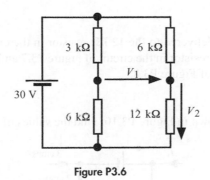

Figure P3.6

Problem 3.7

Find the values of the voltages V_1 and V_2 in the circuit of Figure P3.7.

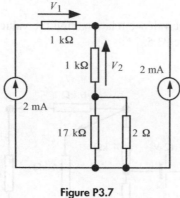

Figure P3.7

Problem 3.8

Find the values of the voltages V_1 and V_2 in the circuit of Figure P3.8.

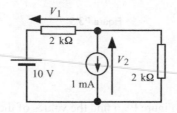

Figure P3.8

Problem 3.9

Calculate the power delivered to the 12 kΩ resistor in the circuit of Figure P3.6, the 'horizontal' 1 kΩ resistor in the circuit of Figure P3.7 and the right-hand 2 kΩ resistor in the circuit of Figure P3.8.

Problem 3.10

Part of a circuit is shown in Figure P3.10. Find the values of I_1, I_2 and V.

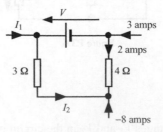

Figure P3.10

Problem 3.11

In the circuit shown in Figure P3.11 device X requires 4 V at 1.5mA and device Y operates at 2 V and 1 mA. The two devices are to be operated from a single 9 V battery as shown. Design the circuit – in other words, specify appropriate values of R_1 and R_2.

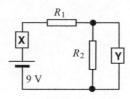

Figure P3.11

4

Circuit Analysis

In the exercises of the previous chapter our analysis was eased by the fact that the circuits were simple, so that spotting the correct sequence in which to apply circuit laws and equivalences was not too difficult. But in real life circuits are *not* simple, and spotting the correct approach to analysis has much in common with solving a fiendish SuDoKu puzzle or *The Times* crossword. What we need (as human beings) is a systematic approach that is foolproof and can be applied mechanically. If a computer is to be used to predict circuit performance, as is often the case, then it too requires a systematic approach to circuit analysis. The principal systematic approach is called *nodal analysis*: it is described below and illustrated by application to the circuit of Figure 4.1.

4.1 Nodal Analysis

Consider the circuit of Figure 4.1, a circuit which is sufficiently complicated that it would be extremely difficult to analyse in the *ad hoc* manner we adopted in Chapter 3.

We first establish how many different voltages there may be in the circuit – and therefore how many we must find by analysis – by removing the components and leaving the connections (Figure 4.2). Since connecting wires are assumed to have zero resistance (so that the voltage at any point along them is the same), we use the shading shown in Figure 4.2 to identify areas of constant voltage. We call these areas 'nodes'. We see from Figure 4.2 that for the circuit of Figure 4.1 there are four different parts, each of which can have a unique voltage. If we connect one probe of a voltmeter to one of these nodes (Figure 4.3) there are three different

Introductory Circuits   Robert Spence
© 2008 John Wiley & Sons, Ltd

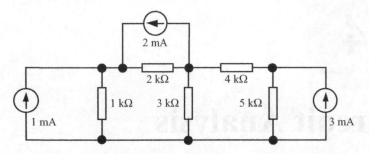

Figure 4.1   The circuit to be analysed

voltages that can be measured in the circuit by the other probe: Figure 4.3 shows one of those voltages being measured.

If we want to talk about the voltages in the circuit we need some reference for voltage, otherwise the statement 'the voltage at this point is 4 V' will be met with the question '4 V with respect to what?' So, mindful of Figure 4.3, we choose one node of the circuit to have zero voltage. We indicate our choice of voltage reference point either by connecting an 'earth' symbol as shown in Figure 4.4 or by labelling the node '0 V'. This is the *first* of only *four steps* involved in systematic circuit analysis.

We then label the remaining three nodes (e.g., A, B and C in Figure 4.4) to indicate which voltages – measured with respect to the reference node – must be found by analysis. In labelling the remaining nodes we have indicated which voltages remain to be calculated. Indeed, to remind ourselves that it is the voltage between a node and the reference node that is of interest, we use an arrow to indicate the voltage reference direction and a label (e.g., V_A) to give a name to that

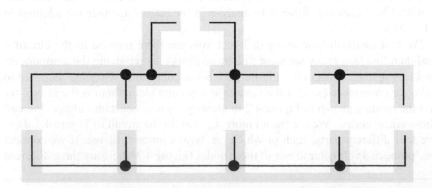

Figure 4.2   The identification of circuit nodes for the circuit of Figure 4.1

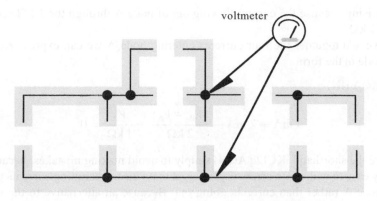

Figure 4.3   The measurement of a voltage

voltage. We have now completed the *second* of the four steps involved in circuit analysis. We should note that the three voltages we have identified in Figure 4.4 will not necessarily be positive in value in the reference directions shown.

The third step in systematic circuit analysis involves the application of KCL at the nodes (here A, B and C) associated with the as-yet-unknown voltages V_A, V_B and V_C. Let us take node A first. There are four components connected to A and therefore four currents to add up and set equal to zero. We shall arbitrarily choose to sum the currents flowing *into* node A. Two of the four currents are supplied by current sources: their values are 1 and 2 mA. By Ohm's law the current entering node A via the 2 kΩ resistor is the voltage across that resistor ($V_B - V_A$) divided by 2 kΩ. The current entering node A via the 1 kΩ resistor is $-V_A/1$ kΩ, the minus

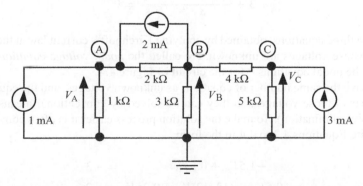

Figure 4.4   The selection of a voltage reference point and the identification of other nodal voltages

sign arising because the current flowing *out* of node A through the 1 kΩ resistor is $V_A/1$ kΩ.

If we add together the four currents entering node A we can express KCL at that node in the form

KCL @ A (IN)

$$1\,\text{mA} + 2\,\text{mA} + \frac{(V_B - V_A)}{2\,\text{k}\Omega} - \frac{V_A}{1\,\text{k}\Omega} = 0 \qquad (4.1)$$

We use the shorthand KCL@A(IN) simply to avoid making mistakes, because it is very easy to make sign errors if you forget that you are summing currents going into node A rather than currents going out. Because an alternative to the set of units [A, V, Ω] is [mA, V, kΩ] we can rewrite Equation (4.1) as

$$1 + 2 + \frac{(V_B - V_A)}{2} - V_A = 0 \qquad (4.2)$$

We next express KCL for node B, again choosing to sum currents flowing into that node:

KCL @ B (IN)

$$-2 + \frac{(V_A - V_B)}{2} + \frac{(V_C - V_B)}{4} - \frac{V_B}{3} = 0 \qquad (4.3)$$

Finally, we apply KCL at node C to obtain:

KCL @ C (IN)

$$3 + \frac{(V_B - V_C)}{4} - \frac{V_C}{5} = 0 \qquad (4.4)$$

These three equations, obtained by applying Kirchhoff's current law at the three nodes whose voltages are unknown, are called the *nodal voltage equations* – or simply the nodal equations – for the circuit of Figure 4.4.

We have the same number of equations as unknown voltages, and the equations are linear in those voltages, so they can be solved by conventional means (e.g., successive elimination). To make the solution process easier it is often convenient to rewrite Equations 4.2 to 4.4 in the form:

$$
\begin{aligned}
-1.5V_A + 0.5V_B &= -3 \\
0.5V_A - (13/12)V_B + 0.25V_C &= 2 \\
0.25V_B - 0.45V_C &= -3
\end{aligned} \qquad (4.5)
$$

In the fourth step of systematic analysis, successive elimination applied to this set of equations will show that

$$V_A = 16/7 \text{ V}, \quad V_B = 6/7 \text{ V} \quad \text{and} \quad V_C = 50/7 \text{ V}$$

It is important to note that once we know the values of V_A, V_B and V_C we can easily find all the component voltages, and hence currents, in the circuit. For example (see Figure 4.4), the current flowing to the left in the 4 kΩ resistor is

$$(V_C - V_B)/4 = (50/7 - 6/7)/4 = 11/7 \text{ mA}$$

Voltage sources

The systematic analysis of a circuit becomes a little more difficult when voltage sources are present in the circuit, although the four steps involved are the same. To illustrate the procedure involved we select the circuit of Figure 4.5 as an example.

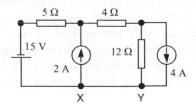

Figure 4.5 A circuit containing a voltage source

In order to avoid any confusion regarding the two connection points X and Y which, of course, are at the same voltage, it can be helpful (Figure 4.6) to redraw the circuit and use a single connection point to represent the node. As with the previous example we first select, arbitrarily, a voltage reference node: this is indicated by the earth symbol in Figure 4.7. We then label the other nodes W, A and B, as shown in Figure 4.7.

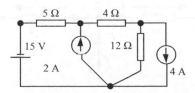

Figure 4.6 The circuit of Figure 4.5 redrawn to emphasize the circuit nodes

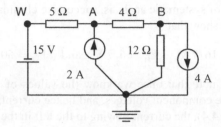

Figure 4.7   A voltage reference node has been chosen and other circuit nodes labelled

At this point we notice that the voltage V_W is already specified as -15 V: we therefore do not need to solve equations to find it. The voltages V_A and V_B are the 'unknowns' whose value must be found by applying KCL at nodes A and B and solving the resulting nodal equations.

Let us take node A first. There are three components connected to A and therefore three currents to add up and set equal to zero. We shall arbitrarily choose to sum the currents flowing *into* node A. By Ohm's law the current (labelled I_1 in Figure 4.8) entering node A via the 5 Ω resistor is the voltage across that resistor $(-15 - V_A)$ divided by 5 Ω. The current I_2 entering node A via the current source is simply 2 A. The third current I_3, entering node A via the 4 Ω resistor, is again obtained from Ohm's law. The voltage across the 4 Ω resistor in the reference direction shown is $V_B - V_A$ so that the current is $(V_B - V_A)/4$. If we add I_1, I_2 and I_3 together we can express KCL at node A in the form

KCL @ A (IN)

$$\frac{(-15 - V_A)}{5} + 2 + \frac{(V_B - V_A)}{4} = 0 \tag{4.6}$$

We then express KCL for node B though arbitrarily choosing to sum currents flowing out of B:

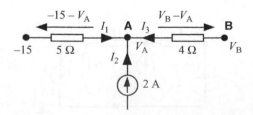

Figure 4.8   Illustration of the application of KCL at node A of the circuit of Figure 4.7

KCL @ B (OUT)

$$\frac{(V_B - V_A)}{4} + \frac{V_B}{12} + 4 = 0 \tag{4.7}$$

Equations (4.6) and (4.7) are the nodal equations for the circuit of Figure 4.7. By simple elimination we obtain V_A and V_B:

$$V_A = -320/21 \text{ V}, \quad V_B = -164/7 \text{ V}$$

Once we know V_A and V_B we can easily find all the component voltages, and hence currents, in the circuit. For example (see Figure 4.7), the current flowing to the left in the 4 Ω resistor is $(V_B - V_A)/4 = (-164/7 + 320/21)/4 = -2.05$ A.

Example 4.1

There are situations in which the choice of unknown nodal voltages may not be obvious. Let us, for example, work with the same circuit as before (Figure 4.7), but assume that the node previously labelled A has been chosen as the reference node, as indicated in Figure 4.9. This is a perfectly valid choice of reference node, but the nodes whose voltages with respect to the reference are unknown need to be identified with care. We might first choose node X because no voltage source is connected directly to it, thereby identifying V_X as an unknown voltage. We could also choose node Y since there is no direct connection to the reference node via a voltage source: V_Y is the voltage at node Y with respect to the reference node. But what about node Z? The voltage V_Z is not an unknown voltage because, once V_Y is known, so is V_Z because $V_Z = V_Y - 15$. Thus, having identified the unknown nodal voltages V_X and V_Y, systematic analysis of the circuit of Figure 4.5 for the choice of reference node shown in Figure 4.9 would now proceed by applying

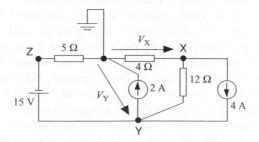

Figure 4.9 Preparation for the analysis of the circuit of Figure 4.7, using a different choice of reference node

KCL at nodes X and Y to obtain two linear equations (nodal voltage equations) in V_X and V_Y which can easily be solved. The results will be identical with the earlier analysis except, of course, that $V_X - V_Y$ will have the same value as V_B and V_Y will be equal to $-V_A$.

4.2 Superposition

Systematic circuit analysis, leading to nodal voltage equations, is not the only method of analysis, even though it has advantages. Another method which may offer an attractive alternative can be illustrated by examining the numerical terms on the right-hand sides of Equations 4.5, the nodal voltage equations describing the circuit of Figure 4.4. The term −3 on the right hand side in the first of the three equations is due to the addition, at node A, of the 1 and 2 mA sources. The term 2 in the second equation is due to the 2 mA source connected to node B. And the −3 term in the third equation in (4.5) is due to the 3 mA source connected to node C. If you solve the nodal voltage equations (4.5) by multiplying each equation by an appropriate constant and adding the results (i.e., by successive elimination) you will find that V_A, for example, is directly proportional to each of the sources contributing to the right-hand sides of (4.5). In other words, we could express V_A in the form

$$V_A = a(1 \text{ mA}) + b(2 \text{ mA}) + c(3 \text{ mA}) \tag{4.8}$$

where a, b and c are constants. For the moment it does not matter what the values of these constants are. What *is* significant is that Equation (4.8) shows that we can calculate V_A by means of three simpler analyses. In the first (Figure 4.10a) we replace the 2 mA source by an open-circuit (so that b is multiplied by zero) and the 3 mA source also by an open-circuit (so that c is multiplied by zero), creating a circuit we can analyse to find that component of V_A due to the 1 mA source. We then analyse another circuit (Figure 4.10b) in which the 1 mA source in the original is replaced by an open-circuit (so that a is multiplied by zero) and the 3 mA source by an open-circuit (so that c is multiplied by zero), creating another circuit we can solve to find V_A, the V_A due solely to the 2 mA source. We repeat this process to find the value of V_A due to the 3 mA source. The principle of superposition now enables us to add the three separately calculated values of V_A together to find the actual value of V_A in the circuit of Figure 4.4.

As stated, this method of circuit analysis is based on the *superposition principle*, which states that *in any system in which there is a linear relation between sources and responses, the response of a system to a number of simultaneous sources is the sum of the responses to each source applied separately*. In our example the sources were current sources and the response was the voltage V_A.

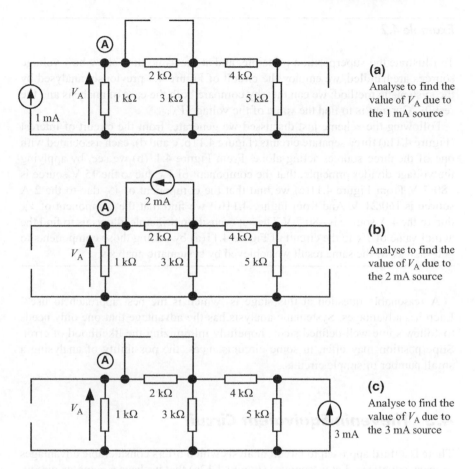

(a)
Analyse to find the
value of V_A due to
the 1 mA source

(b)
Analyse to find the
value of V_A due to
the 2 mA source

(c)
Analyse to find the
value of V_A due to
the 3 mA source

Figure 4.10   Illustration of the use of the superposition principle to find the voltage V_A at node A in the circuit of Figure 4.4. The three calculated voltages are added together to find the actual value of V_A.

Example 4.2

To illustrate the superposition principle, and at the same time show how voltage sources are handled, we employ the circuit of Figure 4.7 previously analysed by the systematic method: we can thereby compare both the ease of analysis and the results. Our aim is to find the value of the voltage V_A.

Following the scheme just discussed we generate, from the circuit of interest (Figure 4.11a) three separate circuits (Figure 4.11b, c and d), each associated with one of the three sources acting alone. From Figure 4.11(b) we see, by applying the voltage divider principle, that the component of V_A due to the 15 V source is $-80/7$ V. From Figure 4.11(c) we find that the component of V_A due to the 2 A source is $160/21$ V. And from Figure 4.11(d) we find that the component of V_A due to the 4 A source is $-80/7$ V. The superposition principle allows us to find the actual value of V_A in the circuit of Figure 4.11(a) by adding those components to get $-320/21$ V, the same result we achieved by systematic analysis.

A reasonable question at this stage is 'which is the best approach to use?' Each has advantages. Systematic analysis has the advantage that one only needs to follow some well-defined steps, hopefully minimizing the likelihood of error. Superposition may offer, in some circumstances, the possibility of analysing a small number of simple circuits.

4.3   Thevenin Equivalent Circuit

There is a third approach to circuit analysis which offers considerable advantages in many situations. Let us suppose (Figure 4.12a) that we have a complex circuit, here denoted N, which is connected by its two external terminals to a simple external circuit, here a single resistor of value R. N might be the HiFi amplifier that lets you listen to music, or an operational amplifier of the sort treated later in Chapter 6. In the first case the resistor R might represent a loudspeaker; in the second it might be a single resistor. In both cases our principal interest may be in the voltage V and current I associated with the resistor R, and not at all with the voltages and currents internal to N. Now imagine that we need to know the voltage V across R as we vary R over a range of (say) 20 values. To find those voltages we could use systematic analysis 20 times, but that would take a great deal of effort. Fortunately a theorem due to Thevenin, a French engineer, drastically reduces the effort involved by allowing us to represent the circuit N by (Figure 4.12b) a voltage source V_{OC} in series with a resistor R_O. Then, the voltage V across the load R can easily be found (e.g., by voltage divider action) even if we have to repeat that

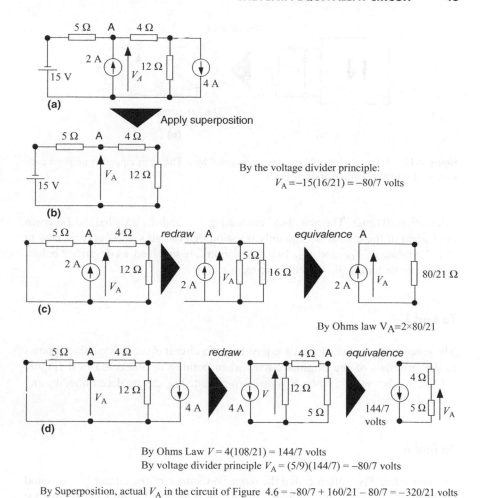

By the voltage divider principle:
$$V_A = -15(16/21) = -80/7 \text{ volts}$$

By Ohms law $V_A = 2 \times 80/21$

By Ohms Law $V = 4(108/21) = 144/7$ volts
By voltage divider principle $V_A = (5/9)(144/7) = -80/7$ volts

By Superposition, actual V_A in the circuit of Figure 4.6 $= -80/7 + 160/21 - 80/7 = -320/21$ volts

Figure 4.11 Illustrating the use of superposition to analyse circuit behaviour

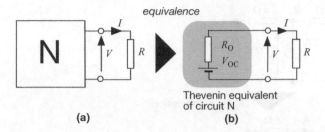

Figure 4.12   A linear circuit N can be represented by a Thevenin equivalent circuit consisting of a voltage source and a resistor

calculation 20 times. The new 'box' containing V_{OC} and R_O is called the *Thevenin equivalent* of the circuit N. The only question that remains, of course, is how to find the values of V_{OC} and R_O. Two calculations are involved, as described below, and are immediately illustrated by an example.

To find V_{OC}

The voltage V_{OC} in the Thevenin equivalent of a circuit is the voltage that appears at the terminals of the original circuit when nothing is connected to it (Figure 4.13a): in other words, when N is on open-circuit. V_{OC} can be determined by any method of analysis.

To find R_O

The resistance R_O – often called the *output resistance* of the circuit N – is found by creating a new circuit (Figure 4.13b), which is identical with N except that we assign zero values to all independent sources. In other words we replace any ideal voltage source by a short-circuit and any ideal current source by an open-circuit. The value of R_O is then the resistance between the two external terminals of this new circuit.

The computational effort required to find V_{OC} and R_O is roughly equivalent to that associated with two circuit analyses, so if the voltage across R in Figure 4.12(a) is to be found for four or more values of R a considerable saving of effort is possible. The Thevenin equivalent circuit in Figure 4.12b is in a real sense a model of the circuit N, so we may often refer to a *Thevenin model* of a circuit. To clarify the calculation of a Thevenin model we use a simple circuit for illustration.

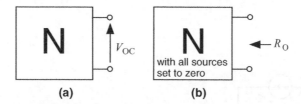

(a)   **(b)**

Figure 4.13 Calculations required to find the two parameters defining a Thevenin equivalent circuit

---

Example 4.3

So that we can check our example we use the same circuit (see Figure 4.7) employed in Examples 4.1 and 4.2, and assume that the circuit whose Thevenin model we require is that shown shaded in Figure 4.14(a). In other words the 4 Ω resistor is the resistance R of Figure 4.12(a), and we are interested only in the voltage across that resistor, not in any other voltage or current inside the shaded region.

We first calculate the voltage V_{OC} using the circuit of Figure 4.14(b). It is not necessary to undertake a nodal analysis because we can see that: (a) the current of 4 A flows solely in the 12 Ω resistor, setting up a voltage of 48 V; and that (b) the 2 A source creates, across the 5 Ω resistor, a voltage of 10 V which, when added to -15 V provides a voltage of -5 V. The addition of 48 and -5 V yields an open-circuit voltage V_{OC} of 43 V.

To find R_O we replace the current sources by open-circuits and the voltage source by a short-circuit to generate the circuit of Figure 4.14(c). The resistance between the two external terminals is easily seen to be 17 Ω.

These two calculations, of V_{OC} and R_O, now allow us to represent the circuit of Figure 4.14(a) by the Thevenin model of Figure 4.14(d). If we now reconnect the 4 Ω resistor (Figure 4.14d) we can easily find the current through it in the direction from right to left:

$$I = -V_{OC}/(R_O + 4) = -43/21 = -2.05 \text{ A}$$

This result checks with the systematic analysis carried out earlier and leading to Equations (4.6) and (4.7), from which we calculated the current through the 4 Ω resistor to be -2.05 A.

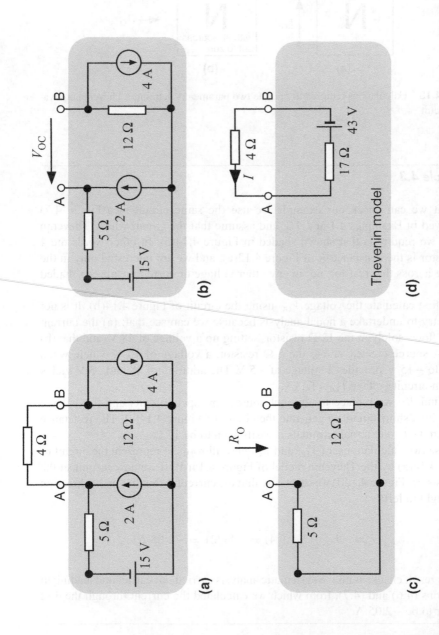

Figure 4.14 Illustration of the development and use of a Thevenin equivalent circuit (a) The circuit of Figure 4.7 redrawn to identify the circuit whose Thevenin model is sought; (b) calculation of the Thevenin parameter V_{OC}; (c) calculation of the Thevenin parameter R_O; (d) use of the Thevenin model to calculate the current through the 4 ohm resistor in the circuit of Figure 4.7(a).

To summarize, the use of a Thevenin model is appropriate when the internal behaviour of a circuit is of no interest and only one voltage or current in an external circuit is of concern.

4.4 Norton Equivalent Circuit

A useful alternative to the Thevenin model is the Norton equivalent circuit. Just as a two-terminal circuit composed of linear resistors and sources can be modelled by an ideal voltage source in series with a resistor (Figure 4.15a), it can also be modelled by an ideal current source in parallel with a resistor (Figure 4.15b). The two models are, in fact, simply related, making their use in circuit analysis rather flexible.

Given the parameters V_{OC} and R_O of a Thevenin model, it is a simple matter to find the parameters I_{SC} and R of the equivalent Norton model. The ideal current source in the Norton model is denoted by I_{SC} because it is the current that would flow in a short-circuit connected between the two terminals. If we connect a short-circuit between the terminals of the Thevenin model of Figure 4.15(a) the current that flows is found, by Ohm's law, to be V_{OC}/R_O. Thus, if the Norton and Thevenin models are equivalent,

$$I_{SC} = V_{OC}/R_O \tag{4.9}$$

We can find the value of R in the Norton model by equating the open-circuit voltage of each model, V_{OC} for the Thevenin model and $I_{SC}R$ for the Norton model. Thus:

$$V_{OC} = I_{SC}R \tag{4.10}$$

Comparison of Equations (4.9) and (4.10) shows that

$$R = R_O \tag{4.11}$$

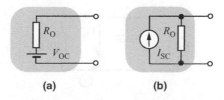

(a)   (b)

Figure 4.15 The Thevenin (a) and Norton (b) equivalent circuits that can model a linear circuit

Conversion from one model to the other is therefore straightforward: the resistances are identical, and Equation (4.9) allows one source value to be derived from the other.

4.5 Problems
Systematic circuit analysis

Problem 4.1

For the circuit of Figure P4.1, and for the indicated choice of voltage reference node, apply KCL at node A. Hence find the value of the voltage V_A. Does your answer agree with the simple application of Ohm's law to the current and voltage of the resistor?

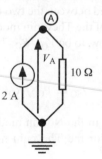

Figure P4.1

Problem 4.2

For the circuit of Figure P4.2, and for the indicated voltage reference node, use nodal analysis to find the voltage V_B.

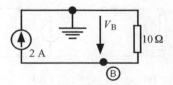

Figure P4.2

Problem 4.3

For the circuit shown in Figure P4.3 a reference node has been chosen. Apply KCL at nodes A and B to obtain the nodal voltage equations. Solve these to find:

(1) V_A and V_B (show them on the circuit diagram)

(2) The voltage across the 2 kΩ resistor

(3) The currents in all resistors

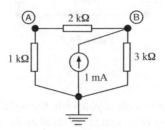

Figure P4.3

Now check that the currents entering each node obey Kirchhoff's current law.

Problem 4.4

Using the same circuit and choice of reference node as in Figure P4.3 apply KCL at node A and at the reference node. Do you obtain two nodal voltage equations in V_A and V_B that can be solved to find these two voltages?

Problem 4.5

Redraw the circuit of Figure P4.3 by combining the series connection of the 1 kΩ and 2 kΩ resistors into a single equivalent resistor. How many unknown voltages are there now? Write down the nodal voltage equation for this new circuit and solve it to find the single nodal voltage. Does it agree with the value of V_B you found in Problem 4 3? Now use the voltage divider principle to find the voltage across the 1 kΩ resistor. Does that agree with the value of V_A found in Problem 4.3?

Problem 4.6

For the circuit of Problem 4.3 (Figure P4.3) choose a different reference node and find the new nodal equations. Solve them and check that the result agrees with the result of Problem 4.3.

Problem 4.7

Choose a reference node for the circuit of Figure P4.7 and obtain the corresponding set of nodal voltage equations.

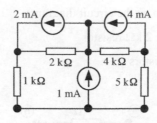

Figure P4.7

Problem 4.8

Write down, but do not solve unless you wish, the nodal voltage equations for the circuit of Figure P4.8. Any node can be selected as the reference node.

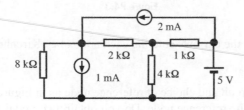

Figure P4.8

Superposition principle

Problem 4.9

For the circuit shown in Figure P4.9 use the superposition principle to find the value of the voltage V.

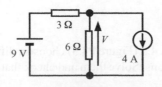

Figure P4.9

Problem 4.10

Use the superposition principle to find the value of the voltage across the 12 kΩ resistor in the circuit of Figure P4.10. Note the crossover, indicating an absence of connection, near the centre. Also note that one resistor has intentionally not been assigned a value!

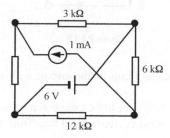

Figure P4.10

Problem 4.11

For the circuit of Figure P4.11 find *one* value of the current I^* for which the current I will fall between the limits of 0.4 and 0.6 mA.

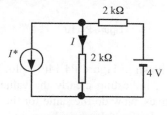

Figure P4.11

Thevenin and Norton equivalent circuits

Problem 4.12

For the circuit within the grey box in Figure P4.12 find both the Thevenin and Norton equivalent circuits. If a 1.6 kΩ resistor were to be connected between A and B what current would flow through it? Perform this calculation using both equivalent circuits.

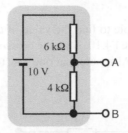

Figure P4.12

Problem 4.13

Find the Thevenin equivalent circuit of the circuit within the grey box in Figure P4.13.

If a voltage source of 12 V were to be connected between A and B, making A positive with respect to B, what current would flow from that source into terminal A?

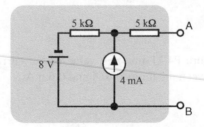

Figure P4.13

Problem 4.14

For the Thevenin model shown in Figure P4.14(a), plot the relation between the external voltages V and I, indicating clearly the value of the intersections on both current and voltage axes. Now do the same for the Norton model of Figure P4.14(b) using the same scales for voltage and current.

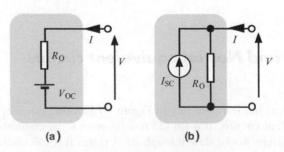

Figure P4.14

5

Controlled Sources and Nonlinear Components

There is another type of electrical component with which we must be familiar if we are to understand – and help to design – the rich collection of functions that circuits can perform. They are known as *controlled sources*. To select the specific example we shall be concerned with in this chapter, they enable a voltage in one part of a circuit to control the current in another part.

5.1 Voltage-controlled Current Source

The symbol for a *voltage-controlled current source* (or VCCS as we shall call it) is shown in Figure 5.1. The VCCS has two pairs of terminals: the voltage across one pair controls the current between the other pair. Figure 5.2 shows the form of the relation between voltage and current, a form that is described by the simple equation

$$I = GV^*$$

(5.1)

where G, which clearly has the dimensions of conductance, is known as the *mutual conductance* to distinguish it from the conductance of a resistor which relates the current through a component directly to the voltage *across the same component*. Note that, as with the independent current source introduced in Chapter 3, the current in a VCCS is independent of the voltage across it, as is evident from Figure 5.2 and Equation (5.1).

Introductory Circuits Robert Spence
© 2008 John Wiley & Sons, Ltd

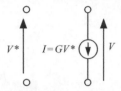

Figure 5.1   Representation of a voltage-controlled current source

We call a VCCS a *dependent source* simply because the current depends for its value upon a voltage elsewhere in the circuit, and to distinguish it from the *independent sources* (the ideal current source and the ideal voltage source) introduced in Chapter 3.

Typically, the idea of a VCCS seems a little unreal: 'What is there between the voltage V^* and the current source in Figure 5.1?' is a question frequently asked. One answer is that the VCCS is a model, but then the question is 'a model of what?'. One device for which the VCCS is a reasonably accurate model is the *transistor* which, of course, appears in its millions in many everyday electronic devices. Though the transistor on its own is not discussed in this book (though it is a vital component of operational amplifiers which are – see Chapters 6, 7 and 8) we show in Figure 5.3 the symbol for a transistor and the measured relation

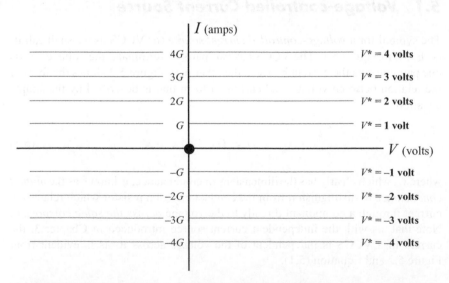

Figure 5.2   The voltage–current characteristics of a voltage-controlled current source

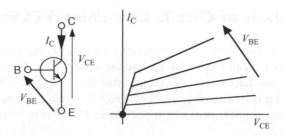

Figure 5.3 The symbol for a transistor, with terminals E (emitter), B (base) and C (collector) and the (sketched) measured characteristics showing the control of the current I_C by the voltage V_{BE}

between what are called the collector current I_C and the base–emitter voltage V_{BE}. Under certain conditions (e.g., with V_{CE} sufficiently large) the characteristics of a transistor approximate to that of a VCCS.

Why do we need the concept of controlled sources? Without them we could not easily model, analyse or design circuits such as amplifiers and switches. Their importance in general is emphasized by the fact that they are so-called active components. We recall from Chapter 3 the discussion of power, and saw that (Figure 5.4a) the power supplied to a component is the product of its voltage V and its current I, provided that the reference directions for V and I are as shown in Figure 5.4a. Since, for a resistor, the signs of V and I are always the same (Figure 5.4b), the resistor cannot be a source of power. However, if we examine the VCCS (Figure 5.4c) we see that the product of V and I can be negative, showing that the VCCS can supply power. For this reason it is classed as an active component.

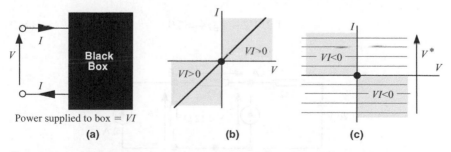

Figure 5.4 (a) The condition for a component to be passive; (b) a resistor is passive: it cannot supply energy; (c) a VCCS is said to be active because it can supply energy if it is operating in the appropriate shaded region

5.2 Analysis of Circuits Containing VCCSs

Because the VCCS is a linear component (see Equation 5.1) the analysis of a circuit containing a VCCS can proceed in exactly the same way as for the DC circuits with which we are familiar, as an example will show. However, the presence of a VCCS does introduce some features of which we have to be aware when, for example, making use of superposition or creating Thevenin or Norton models, as we shall soon see.

Example 5.1

The circuit of Figure 5.5 contains one VCCS as well as two independent sources. Using the systematic method of circuit analysis introduced in Chapter 4 we first select the voltage reference point: this has been indicated by the earth symbol in Figure 5.5. In the second step we identify the location of the unknown voltages: again, these have been indicated by the labels A and B at the nodes whose voltage values with respect to the reference are V_A and V_B, respectively. In the third step we simply apply KCL at these two nodes (here we have (arbitrarily) chosen to sum currents leaving the nodes):

KCL @ A (OUT):

$$\frac{(V_A - 20)}{5} - 0.2(V_A - V_B) + \frac{(V_A - V_B)}{4} = 0$$

Note that we have immediately substituted $V_A - V_B$ for V^* because, in the two nodal equations we expect to obtain, we want only two unknowns. Rearranging the above equation we obtain:

$$1.25V_A - V_B/4 = 20 \tag{5.2}$$

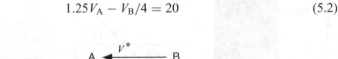

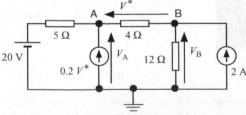

Figure 5.5 The circuit analysed in Example 5.1. It contains one voltage-controlled current source whose mutual conductance is 0.2 A/V (i.e., 0.2 S)

KCL @ B (OUT):

$$-2 + \frac{V_B}{12} + \frac{(V_B - V_A)}{4} = 0$$

which can be arranged in the form

$$-V_A/4 + V_B/3 = 2 \qquad (5.3)$$

Equations (5.2) and (5.3) are the nodal voltage equations for the circuit of Figure 5.5. Because we have two equations in two unknowns we can solve them to obtain V_A ($=344/17$) and V_B ($=360/17$) and, therefore, all the voltages and currents in the circuit.

From Example 5.1 we see that we do not need to modify the systematic method of circuit analysis to cater for the presence of VCCSs. It is useful to note, however, that the VCCS did not contribute to the right-hand side of the nodal equations: the '20' In Equation (5.2) corresponds to the *independent* 20 V source, and the '2' on the right-hand side of Equation (5.3) arises from the *independent* 2 A source. It is because the VCCS is a *dependent* source that it contributes only to the left-hand side of the nodal voltage equations.

Certain rules have to be followed when either superposition is used to analyse a circuit or a Thevenin or Norton model has to be derived, as we discuss below.

Superposition

When using the principle of superposition only the independent sources should be taken in turn as the single source within the circuit. The dependent VCCS should be 'left alone' and treated as part of the circuit to which each source is applied. Some idea of the reason for this rule can be gleaned from the fact that, as Example 5.1 has demonstrated, the VCCS does not contribute to the right-hand side of the nodal equations: it was on the basis of Equation (4.8) that we saw, in Example 4.2 that, for the circuit of Figure 4.6, the value of V_A could be derived by considering, separately, each source which contributes to the right-hand side of the nodal equations. The example below will show how superposition can be applied to a circuit containing a VCCS.

Example 5.2

For each of the two independent sources in the circuit of Figure 5.5 we have created, in Figure 5.6, the circuits needed to apply the superposition principle.

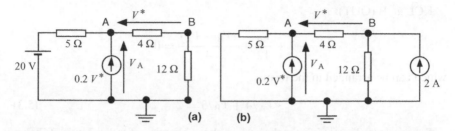

(a)   (b)

Figure 5.6 Circuits prepared for application of the superposition principle to the analysis of the circuit of Figure 5.5. From (a) it is found that the component of V_A due to the independent 20 V source is 320/17 volts. From (b) it is found that the component of V_A due to the independent 2 A source is 24/17 V. Superposition allows us to state that the actual value of V_A in the circuit of Figure 5.5 is $320/17 + 24/17 = 344/17$ V

Systematic circuit analysis shows that for circuit (a) the component of V_A due to the 20 V source is 320/17 V. The analysis of circuit (b) shows that the component of V_A due to the 2 A source is 24/17 V. Superposition then allows us to say that the actual value of V_A is $320/17 + 24/17 = 344/17$ V, which agrees with the value calculated in Example 5.1.

Thevenin models

The difference in treatment of dependent and independent sources also extends to the development of Thevenin models. To calculate the parameter V_{OC} of a Thevenin model we simply carry out a conventional circuit analysis. However, when R_O is to be calculated, and all independent sources are set to zero, *dependent sources such as VCCSs are left untouched.*

In Example 5.3 below we demonstrate the derivation of a Thevenin model for a circuit containing a VCCS. We again select the circuit of Figure 5.5 for illustration so that a comparison can be made with Examples 5.1 and 5.2.

Example 5.3

We assume in this example that, for the circuit of Figure 5.5, the Thevenin model being sought is that describing the circuit connected to the 12 Ω resistor: in other words, we seek the Thevenin model of the circuit within the grey area as shown in Figure 5.7.

We proceed, first, to find the open-circuit voltage V_{OC}, simply by analysing the circuit (a) shown in Figure 5.8. It is found by nodal analysis to be 30 V.

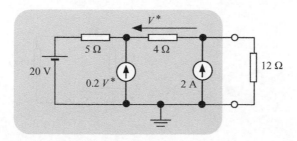

Figure 5.7 The circuit of Figure 5.5 rearranged to show that part for which a Thevenin model is to be found

We now find the parameter R_O by (b) setting to zero all *independent* sources within the circuit to be modelled, but leaving the VCCS untouched. To find the resistance R_O between the external terminals we apply a current of 1 A and calculate the resulting voltage V. Analysis reveals that the voltage V is 5 V. The combination of a current of 1 A and a voltage of 5 V identifies a resistance of 5 Ω between the two terminals.

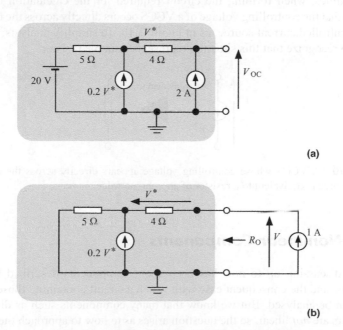

Figure 5.8 The two analyses required to find the Thevenin parameters V_{OC} and R_O of the circuit bounded by the grey box in Figure 5.7 (a) V_{OC} is found by analysis to be 30 V; (b) $R_O = V/1$ A is found by analysis to be 5 Ω

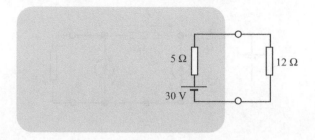

Figure 5.9 Use of the Thevenin model of the circuit of Figure 5.7 to calculate the voltage across the 12 Ω resistor. If the 12 Ω resistor is replaced by another resistor having a different value only a very simple calculation is involved, invoking Ohm's law

As shown in Figure 5.9 we can now use the simple Thevenin model we have derived to calculate the voltage across the 12 Ω to be 30 (12/17) V = 360/17 V, a result which agrees with the calculations carried out in Examples 5.1 and 5.2.

Sometimes, when forming the circuit required for the calculation of R_O, it happens that the controlling voltage of a VCCS occurs directly across the terminals of the controlled current source, as in Figure 5.10. To simplify analysis, it can be useful to recognize that this situation represents a simple resistor!

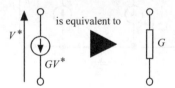

Figure 5.10 A VCCS whose controlling voltage appears directly across the controlled current source is equivalent to a resistor of appropriate value

5.3 Nonlinear Components

All our discussion up to now has been about components described by linear equations, and the consequent ease with which a circuit containing those components can be analysed. But we know that many components such as diodes and transistors are *not* linear, so the question arises as to how to approach the analysis of circuits containing those devices. Certainly, all the many advantages that accrue from linearity are now gone if even a single nonlinear component is present: a new approach to analysis must be found.

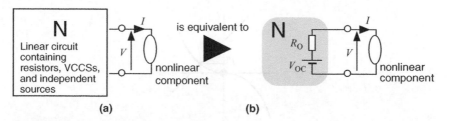

(a)   **(b)**

Figure 5.11   (a) A circuit containing one nonlinear component whose voltage and current are of interest; (b) a representation of the linear part of the circuit by a Thevenin model

Load-line construction

To illustrate an approach that is useful if there is one nonlinear component in a circuit we represent the circuit as shown in Figure 5.11(a), with the nonlinear component connected externally to a circuit (N) that is linear. However, we know that a linear circuit can be represented by a Thevenin model, so we reformulate the problem as shown in Figure 5.11(b).

To analyse the circuit of Figure 5.11(b) we split it into two parts, as shown in Figure 5.12. Consider the right-hand part first, the nonlinear component. A plot of its current- voltage characteristic is shown. Now consider the left-hand part, the circuit N. When the two parts are reconnected to form the circuit of Figure 5.11(a) the voltage V and the current I will be common to both parts, and therefore we designate the voltage of N to be V and its current to be I, *but in the same*

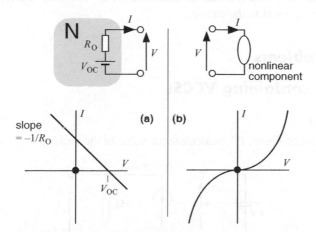

Figure 5.12   Separation of the linear and nonlinear parts of the circuit in Figure 5.11, and their characterization by current~voltage plots

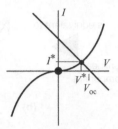

Figure 5.13   The point I, V must lie on both characteristics, and hence at their intersection (I^*, V^*)

direction as with the nonlinear component: this is important. The current~voltage characteristic of N, using the chosen reference directions for V and I, is as shown in Figure 5.12(a) and, as expected, is linear.

Connection of the two parts to obtain the circuit of Figure 5.11(b) has the following consequence. The point (I, V) describing the nonlinear component must lie on the nonlinear characteristic shown in Figure 5.12(b), but the *same* point (I, V) describing the circuit N *also* lies on the linear characteristic shown in Figure 5.12(a): it must therefore occur at the intersection of the two characteristics, as shown in Figure 5.13. The intersection identifies the current I^* and the voltage V^* which are the values of I and V in the original circuit of Figure 5.11(b). Thus, by using the same axes to draw the V–I characteristics of the two parts of the circuit, we have found the values of voltage and current.

Conventionally, the straight line in Figure 5.11(a) is called a *'load-line'*. The graphical method of analysis demonstrated above is valid only if there is a single nonlinear component in the circuit.

5.4 Problems

Circuits containing VCCSs

Problem 5.1

For the circuit of Figure P5.1 calculate the value of the voltage V.

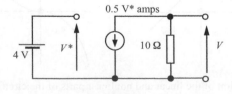

Figure P5.1

Problem 5.2

The circuit of Figure 5.8(a) is reproduced as Figure P5.2. Analyse the circuit to find the value of the voltage V_{OC}.

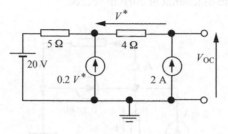

Figure P5.2

Problem 5.3

Choose any value for V^* and calculate the value of the current I in the circuit of Figure P5.3. Would the current be any different in value if the VCCS were to be replaced by a resistor of value 100Ω?

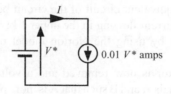

Figure P5.3

Problem 5.4

The circuit of Figure 5.8(b) is reproduced here as Figure P5.4. Find the value of the resistance R_O between the terminals of the grey box.

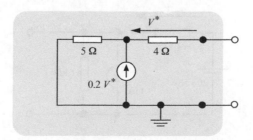

Figure P5.4

Problem 5.5

The circuit shown in Figure P5.5 contains one voltage-controlled current source having a mutual conductance of 0.2 mA per volt, in addition to an independent voltage source and an independent current source.

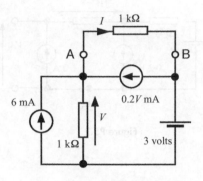

Figure P5.5

By using the superposition principle calculate the value of the current I flowing through the upper 1 kΩ resistor.

Derive the Thévenin equivalent circuit of the circuit below terminals A and B and hence calculate the current flowing in the upper 1 kΩ resistor.

Find the same current by using the Norton model of the circuit below the terminals A and B.

The upper 1 kΩ resistor is now removed and a voltage source of 4.5 V is connected between terminals A and B such that A is more positive than B. Calculate the current that will flow in this voltage source.

Problem 5.6

Find the Thévenin equivalent circuit of the circuit within the grey box in Figure P5.6. The circuit contains a voltage-controlled current source.

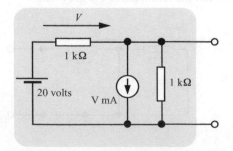

Figure P5.6

Problem 5.7

For the circuit shown in Figure P5.7 find the Thévenin equivalent circuit with respect to terminals A and B. Also, find the corresponding Norton equivalent circuit.

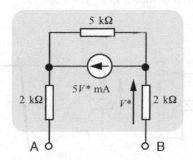

Figure P5.7

Load-line construction

Problem 5.8

The circuit shown in Figure P5.8(a) has application to the generation of a well-regulated voltage supply and is discussed in detail in Chapter 12. The nonlinear component X is described by the current~voltage relation shown in Figure P5.8(b).

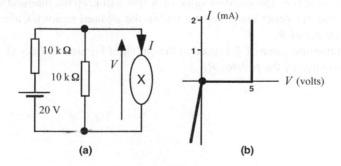

(a) **(b)**

Figure P5.8

By means of a load-line construction find the value of V and I in the circuit of Figure P5.8(a).

If the right-hand 10 kΩ resistor is now replaced by one having a resistance of 5 kΩ, what are the new values of V and I?

Problem 5.9

The circuit of Figure P5.9(a) contains a nonlinear component known as a Zener diode, and forms the basis of a circuit introduced in Chapter 12. A good approximation to its $V\sim I$ relation is shown in Figure P5.9(b).

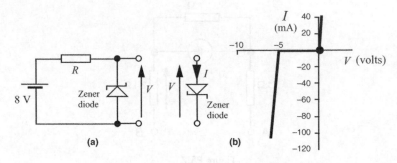

Figure P5.9

A limitation to the performance of this circuit is set by the maximum power rating of the Zener diode which, if exceeded, will lead to irreversible damage due to heating. The maximum power that can be absorbed by the diode and dissipated as heat is 300 mW (i.e., in Figure P5.9b the product VI must be less than or equal to 300 mW).

On the plot provided in Figure P5.9(b) sketch the boundary of permissible power dissipation

Sketch, on the same plot, the load line corresponding to the minimum permissible value of R (i.e., the smallest value of R that will keep the intersection of the load-line and the Zener characteristic within the allowed region). Calculate this minimum value of R.

If the minimum value of R is used in the circuit of Figure P5.9(a), calculate the power dissipated by the resistor R.

Overview: Operational Amplifiers

An extremely valuable, inexpensive and useful component is the operational amplifier, henceforth referred to as the *opamp*. As the name suggests it performs amplification, but the potential it offers is far wider. It is very economical to manufacture, especially in integrated circuit (IC) form.

The variety of operations that the opamp can help to provide is extensive, ranging from amplification to switching and the generation of many different waveforms. In this book we shall examine some of them.

One advantage of the opamp as far as its study and its use in circuit design is concerned is its simplicity. It has two input terminals, and it is the difference in voltage between these terminals that controls a single output voltage. For this reason it is often referred to as a *differential amplifier*. To operate satisfactorily two additional terminals must be connected externally to sources of constant voltage. The current required by the two input terminals is so small (of the order of 10^{-12}A) that, in this book, it is justified to assume these currents to be zero without the risk of serious error in our analyses.

Typically, an opamp is made up from a large number of transistors, resistors and a few capacitors, but this detail will not concern us. We shall view the opamp as a 'black box' described by the relation between its input voltage (the difference between the voltages at its two input terminals) and its output voltage.

Introductory Circuits   Robert Spence
© 2008 John Wiley & Sons, Ltd

6

The Operational Amplifier

In previous chapters we confined our attention to circuits containing linear resistors, independent voltage and current sources and simple nonlinear components, all of which can be approximated by available components. We then extended our investigation to include voltage-controlled current sources which, though ideal, are useful for modelling certain components. We now add another component, the operational amplifier, to the selection available to the circuit designer. Its availability makes possible a very wide range of useful circuits.

6.1 Properties of the Operational Amplifier

The operational amplifier (opamp) has five terminals and can be represented symbolically as shown in Figure 6.1. Two of the terminals, marked with circled plus and minus signs, are connected directly to supply voltages (i.e., essentially constant voltages) which are essential to the correct functioning of the opamp. There are two input terminals between which a voltage V_I is applied and this input voltage controls the output voltage V_O. All the voltages V_O, V^+ and V^- are measured with respect to earth.

If measurements are made and corresponding values of V_I and V_O are noted, the relation between input and output voltages will be of the form shown in Figure 6.2. Over a very small range of V_I – from about -100 mV to $+100$ mV – the relation between input and output voltages is linear and has a high slope – as much as 10^4 to 10^6. Thus, in that region around the origin the opamp is acting as a voltage amplifier with a very high gain. Outside what we shall call this 'linear region' the output voltage is constant (at either V_S or $-V_S$) and unaffected by the input voltage. The existence of two regions in which V_O is constant is useful in digital

Introductory Circuits   Robert Spence
© 2008 John Wiley & Sons, Ltd

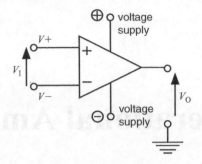

Figure 6.1 The voltages associated with an operational amplifier

circuits based on components exhibiting two states. The voltages $+V_S$ and $-V_S$ are usually within 2 or 3 V of the positive and negative supply voltages.

If, during measurements, the currents entering the input terminals were observed, they would be found to be extremely small – of the order of picoamps. In many applications it can safely be assumed that no current enters these terminals, and that assumption will be made in this book.

The symbolic representation shown in Figure 6.1 will normally be simplified to that shown in Figure 6.3 in which the power supply connections are not shown. It is simply assumed by anyone seeing that representation within a larger circuit diagram that the appropriate power supply connections are intended. The advantage is a simplified circuit diagram. The only disadvantage is that one might be tempted to apply Kirchhoff's current law to the currents entering the three *visible* terminals

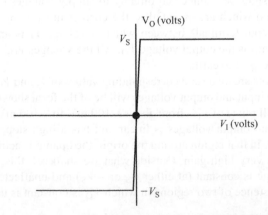

Figure 6.2 The form of the relation between the input V_I and output V_O voltages of an opamp

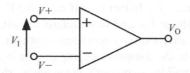

Figure 6.3   The conventional representation of an opamp

in Figure 6.3, in which case it would be erroneously concluded that no current can flow out of the output terminal because none flows into the input terminals. The error comes from ignoring the currents flowing to and from the voltage supplies.

Because, in the central linear region, the output voltage is proportional to the *difference* V_I between the two input voltages V^+ and V^-, the opamp is often referred to as a *differential amplifier*.

In this book we shall regard the operational amplifier as a 'black box': we shall not be concerned with its interior which typically contains as many as 100 or even more components.

6.2   Large-signal Operation

We shall first investigate the behaviour of the opamp when there is no restriction on the values that the input voltage V_I can assume. In the next chapter we shall encounter circuits which exploit, and are restricted to, operation in the linear region.

The comparator

Figure 6.4 shows a circuit which forms the basis of many applications of the opamp. It is called a *comparator*. Recall that when V_I is greater than about $100\,\mu V$ or less than about $-100\,\mu V$, the output voltage V_O is fixed at either its positive or

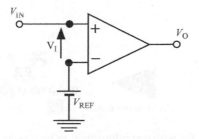

Figure 6.4   The comparator

negative extremes V_S and $-V_S$. In that case if we make V_I the difference between some arbitrary input voltage V_{IN} and some reference voltage V_{REF}, then V_O will be at its positive or negative extreme according to whether V_{IN} is greater or less than V_{REF}. In other words, the comparator compares the input voltage V_{IN} with the reference voltage and provides an output voltage whose sign indicates whether or not V_{IN} is greater than V_{REF}.

In this chapter we look at some applications for which the comparator is the key circuit.

Analog-to-digital (A–D) conversion

We often wish to observe 'real world' quantities that are continuous, such as the temperature inside an engine, the stress in a girder, or the blood pressure of a patient undergoing surgery. However, it is often convenient to process this information digitally (Figure 6.5). We therefore need an analog-to-digital (A-D) converter.

The very simple A–D converter shown in Figure 6.6 illustrates one principle on which such a converter could be based. The chain of four resistors of equal value, and connected to a 4 V supply, ensures that voltages of 1, 2 and 3 V are available to form reference voltages for three comparators. These voltages are connected to the negative inputs of three opamps.

The single (analog) input voltage V_{IN} is applied to the positive input terminals of all three opamps. As an example consider opamp B: its output voltage will be equal to $+V_S$ only if V_{IN} is greater than 2 V, otherwise V_O is equal to $-V_S$. By considering the effect of all possible ranges of V_{IN} between 0 and 4 V we can produce a table (Table 6.1a) relating the three output voltages to the single input voltage. If, for example, the voltages V_S and $-V_S$ correspond to binary 1 and 0, Table 6.1(b) shows the performance of the A–D converter expressed in binary notation. Appropriate logic circuits can operate on the voltages V_A, V_B and V_C as required.

Figure 6.5 It is often convenient for analogue information to be transformed to digital form for easier processing

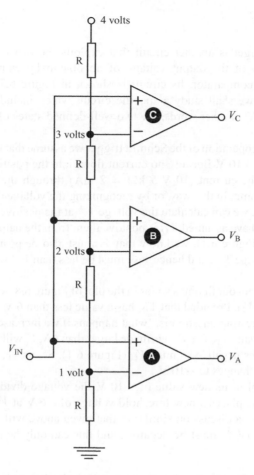

Figure 6.6 A very simple A–D converter based on the use of comparators

Table 6.1 For the A–D converter of Figure 6.6, (a) shows the values of the opamp output voltages for different ranges of the input voltage. In (b) the same outputs are expressed in binary form

V_{IN}	V_A	V_B	V_C		A	B	C
Between 3 and 4 V	V_S	V_S	V_S		1	1	1
Between 2 and 3 V	V_S	V_S	$-V_S$		1	1	0
Between 1 and 2 V	V_S	$-V_S$	$-V_S$		1	0	0
Between 0 and 1 V	$-V_S$	$-V_S$	$-V_S$		0	0	0
(a)					(b)		

Schmitt trigger

The Schmitt trigger is another circuit that exploits the two well-defined states $(+V_S$ and $-V_S)$ of the output voltage of an opamp by employing the basic property of the comparator. Its circuit is shown in Figure 6.7. In order to explain its action we shall study a specific circuit which includes an opamp for which $V_S = 10$ V. In other words, the two well-defined states of V_O are $+10$ and -10 V.

To explain the operation of the Schmitt trigger we assume that the output voltage V_O is initially at $+10$ V. Because no current flows into the positive input terminal of the opamp, the current (10 V/5 k$\Omega = 2$ mA) through the 2 kΩ and 3 kΩ resistors is the same. In this way, or by recognizing the voltage divider formed by the two resistors, we can calculate the voltage V^+ at the positive input terminal to be 6 V. If, as we have assumed, V_O is positive then, from the nature of the opamp's characteristics (Figure 6.1), we know that V_I must also be positive. And if V_I is positive, the voltage V^-, and hence V_{IN}, must be less than V^+ which we have just calculated to be 6 V.

We can represent our findings so far in the plot of Figure 6.8, where V_O is plotted against V_{IN} (*not* V_I). Provided that V_{IN} has a value less than 6 V, V_O will remain at $+10$ V. The interesting question is, 'what happens if we increase V_{IN} above 6 V?'

If V^+ is at 6 V and V_{IN} ($= V^-$) is made larger than 6 V, V_I will become negative, and if V_I becomes negative so must V_O (Figure 6.1). So, as V_{IN} begins to exceed 6 V, V_O quickly changes to -10 V.

With V_O equal to its new value of -10 V, the voltage divider formed by the two resistors now places a new threshold voltage of -6 V at V^+. With this new threshold voltage, a discussion similar to that given above will show that, for V_O to remain at -10V, V_I must be negative, and this can only be achieved if V_{IN} is

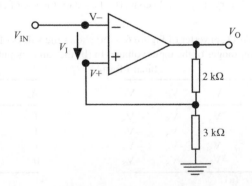

Figure 6.7   The Schmitt trigger

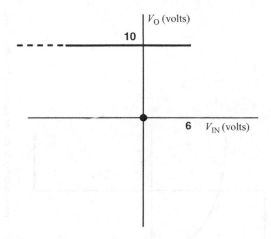

Figure 6.8 The permissible range of V_{IN} in the Schmitt trigger of Figure 6.7 while V_O is at $+10$ V

greater than -6 V. We can now add this detail to the characteristic of Figure 6.8 to get Figure 6.9 which completely defines the behaviour of the Schmitt trigger for any value of V_{IN}. Note, however, that with V_{IN} at any value between -6 and $+6$ V one cannot say whether V_O will be at $+10$ or -10 V; it depends upon *how* V_{IN} *arrived* at that value.

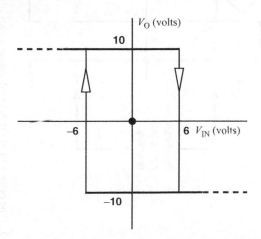

Figure 6.9 The conditions relating the input and output voltages of the Schmitt trigger of Figure 6.7, showing the fast transitions between the possible stable output states Arrows denote a fast transition

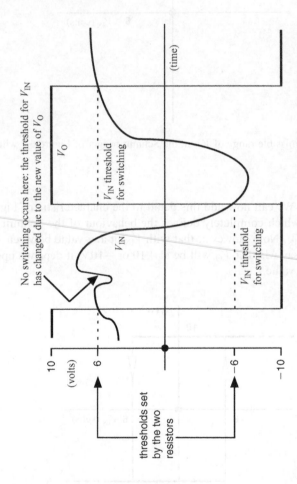

Figure 6.10   The behaviour of the Schmitt trigger for an arbitrary input voltage waveform. Take particular note of the fact that the threshold for V_{IN} changes when the output voltage changes

Example 6.1

To emphasize the fact that the threshold voltage at V^+ changes when the value of V_O changes we examine the behaviour of the Schmitt trigger circuit (Figure 6.7) for the waveform of V_{IN} shown in Figure 6.10. The first time that V_{IN} rises above 6 V the value of V_O changes, but the second time nothing happens because the threshold voltage at V^+ is different.

Although the Schmitt Trigger circuit may at first seem complex it is useful to be reminded of the very simple comparator on whose property it is based.

Another trigger circuit

The circuit shown in Figure 6.11 exhibits behaviour similar to that of the Schmitt trigger. To understand its operation we again assume that the opamp output voltage V_O is at $+10$ V. Because V_O is positive, V_I must be positive. To make V_O negative we must make V_I negative. The only way we can make V_I negative is to alter the value of V_{IN}. To see what value of V_{IN} is needed to make V_I negative we draw the circuit of Figure 6.12 which is that part of the circuit which determines the value of V_I once V_O is fixed. What we have in Figure 6.12 is a circuit that can be treated as a voltage divider because no current is drawn from the central node by its connection to an input terminal of the opamp. From this circuit we see that if V_O is fixed at 10 V then, as V_{IN} is decreased in value, V_I will also decrease. We are interested in when V_I has fallen to zero and is about to go negative (whereupon V_O will change sign), so we analyse the circuit of Figure 6.12 with V_I set to zero.

Since the currents in the resistors R_1 and R_2 are the same we can invoke Ohm's law to write

$$(0 - V_{IN})/R_1 = (10 - 0)/R_2$$
$$\text{so that } V_{IN} = -(R_1/R_2) \times 10V \tag{6.1}$$

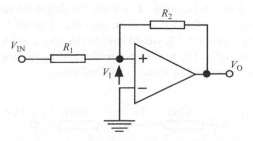

Figure 6.11 Another trigger circuit whose output state can be controlled by an input voltage

$$V_{IN} = ? \quad R_1 \quad V_1 = 0 \quad R_2 \quad V_O = 10 \text{ V}$$

Figure 6.12   Circuit relevant to the calculation of the value of V_{IN} needed to cause the output voltage of the circuit of Figure 6.11 to change from $+10$ to -10 V

If we choose $R_2 = 2R_1$, we find that V_{IN} is equal to -5 V. In this case a reduction of V_{IN} towards -5 will reduce V_1 to zero, and any further slight reduction of V_{IN} will cause V_1 to be negative and V_O to change its value to -10 V. As with the Schmitt trigger, there will now be a new threshold value of V_{IN} required to change V_O back to $+10$ V.

Example 6.2

A voltage V_{IN} having the triangular waveform (but with nonzero average value) shown in Figure 6.13(a) is applied to the input of the circuit shown in Figure 6.13(b). Find the resulting waveform of the voltage V_O. The output voltage V_O of the opamp limits at ± 8 V.

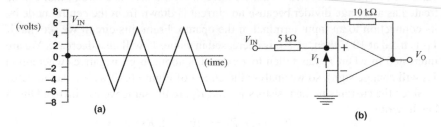

Figure 6.13   The waveform (a) of the input applied to the trigger circuit (b)

The subcircuit shown in Figure 6.14 controls the threshold values of V_{IN}. With V_O at 8 v it is easily seen that for V_I to be zero the value of V_{IN} must be -4 V. Similarly, when V_O is at -8 V, V_{IN} would have to rise above $+4$ V in order to make V_1, and hence V_O, positive. The resulting waveform of V_O is shown in Figure 6.15, with the threshold values of V_{IN} shown as dashed lines.

$$V_{IN} = ? \quad 5 \text{ k}\Omega \quad V_1 = 0 \quad 10 \text{ k}\Omega \quad V_O = 8 \text{ V}$$

Figure 6.14   Model allowing calculation of the threshold voltages for the circuit of Figure 6.13(a)

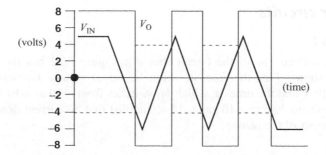

Figure 6.15 The waveform of the output voltage of the trigger circuit of Figure 6.13(a)

6.3 Problems

Opamp as a comparator

Problem 6.1

For each of the circuits shown in Figure P6.1 find the value of the output voltage V_O. Assume that the limits to the value of V_O are $+10$ and -10 V.

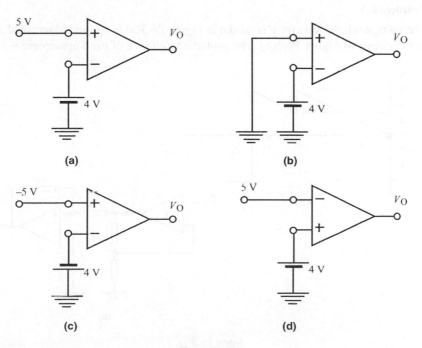

Figure P6.1

Trigger circuits

Problem 6.2

The input voltage V_{IN} of the circuit shown in Figure P6.2 has the triangular waveform shown. Find the times at which the output voltage V_O switches from +10 to −10 V and the times at which V_O switches from −10 to +10 V. Assume that the limits to V_O are +10 and −10 V, and that that no current flows into the input terminals of the opamp.

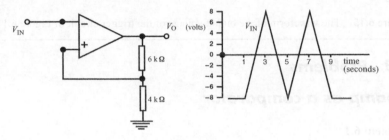

Figure P6.2

Problem 6.3

The voltage whose waveform is shown in Figure P6.3(a) is applied to terminal A in the circuit of Figure P6.3(b). The saturation voltages of each opamp are +15

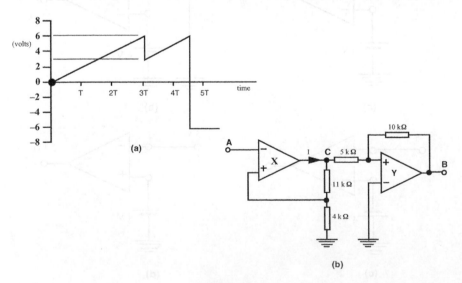

Figure P6.3

and −15 V; otherwise the opamps can be considered ideal. Both opamps have their outputs initially at +15 V.

Provide a dimensioned sketch of the waveform of the voltage at terminals B and C.

Provide a dimensioned sketch of the waveform of the current I.

If the negative input terminal of opamp Y is connected to a voltage source instead of earth, what value must that voltage source have in order to prevent any variation in the voltage at terminal B?

Problem 6.4

Refer to the circuit shown in Figure P6.4. For what range of values of the voltage V will no current flow in the 10 kΩ resistor?

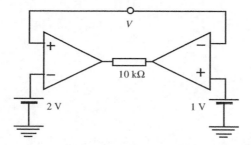

Figure P6.4

Problem 6.5

For the circuit shown in Figure P6.5 find the voltage that V_{IN} must exceed to cause V_O to change from −10 to +10 V, and the voltage that V_{IN} must then fall below to restore V_O to a value of −10V.

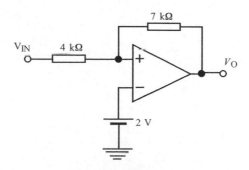

Figure P6.5

and +15 V, otherwise the op amps can be considered ideal. Both op amps have
then power supplies of ±15 V.

Provide a dimensioned sketch of the waveform of the voltage at terminal D
and C.

Provide a dimensioned sketch of the waveform of the current I.

If the negative input terminal of op amp A is connected to another voltage
instead of earth, but still that the voltage source have an effect to give any
variation in the voltage at terminal C?

Problem 6.4

Refer to the circuit shown in the figure below. For what range of values of the voltage
will no current flow in the 10 kΩ resistor?

Figure P6.4

Problem 6.5

For the circuit shown in the figure below, find the voltage that the must exceed in order
to to change from –10 to +10 V, and the value of the V_o, using a diagram below to
produce 10 mA value of –10V.

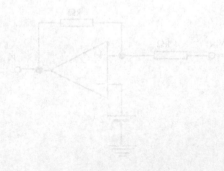

Figure P6.5

7

Linear Operation of the Opamp

In the previous chapter we placed no constraints upon the value of the voltage difference V_I between the two input terminals of an opamp. An advantage was that we could use the two saturated states of an opamp (with $V_O = \pm V_S$) to represent two discrete states of a circuit, with many applications to digital circuits. We now restrict our attention to that part of an opamp's $V_O \sim V_I$ relation close to the origin, where the relation between V_O and V_I is approximately linear and of very high slope (recall Figure 6.1). There are, as we shall see, many opportunities to design useful circuits if the operation of an opamp is maintained within that 'linear region'.

To establish a concept, known as a virtual short-circuit, that considerably simplifies the analysis of such circuits we first examine a circuit called an inverter.

7.1 Virtual Short-circuit

The so-called 'inverter' circuit is shown in Figure 7.1. To determine how it works we first of all need to be aware of relative voltage levels. We already know that the slope of the $V_O \sim V_I$ relation in the high-slope region around the origin can be as high as 10^4 to 10^6. In that case, if V_O lies between the limits $\pm V_S$ of linear operation (say $+10$ and -10 V), the largest value that V_I can exhibit is between 10μV and 1 mV. Usually, the value of V_{IN} is much larger than this, so that it can safely be assumed, for the purpose of analysis, that V_I is zero. As a consequence we say that there is a *virtual short-circuit* between the two input terminals of the opamp. This is different, of course, from there being an actual short-circuit

Introductory Circuits   Robert Spence
© 2008 John Wiley & Sons, Ltd

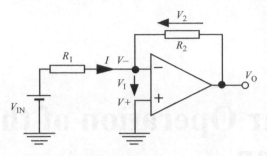

Figure 7.1   The inverter circuit

connected between the input terminals, which would render V_1 and hence V_O zero. The virtual short-circuit is a way of saying that the two input voltages V^+ and V^- are maintained at virtually the same value. It can be useful, when analysing a circuit which exploits the linear region of an opamp, to sketch in by hand a dashed line to remind the analyst that there is negligible difference between the two input voltages. We do that in Example 7.1 below. If, as in the inverter circuit, the positive input terminal is earthed, we say that there is a *virtual earth* at the negative input terminal.

Example 7.1

Because the positive input terminal of the opamp in the circuit of Figure 7.1 is at zero voltage, the principle of the virtual short circuit ensures that V^- is essentially zero. Under these circumstances we say that there is a virtual earth at the negative input terminal: in other words, $V^- = 0$. To remind ourselves of this fact we may redraw the inverter circuit as in Figure 7.2, with a dashed line signifying the negligible difference between the opamp's two input voltages.

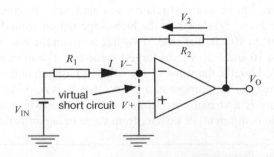

Figure 7.2   The inverter circuit, with the virtual short-circuit sketched in

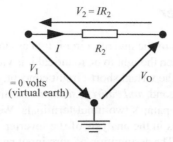

Figure 7.3   An application of KVL to the inverter circuit

The assumption that V^- is zero considerably simplifies analysis. Thus, application of Ohm's law to resistor R_1, the voltage across which is $V_{IN} - V^-$, shows that the current I flowing through it is V_{IN}/R_1. Where does this current flow? Since none can flow into the negative input terminal it all flows through R_2, creating a voltage V_2 equal to $(V_{IN}/R_1)R_2$.

We now apply Kirchhoff's voltage law around the loop (Figure 7.3) which includes the voltages V_O, V_2 and V_1. Taking account of reference directions we obtain

$$V_O + V_2 + V_1 = 0 \tag{7.1}$$

which, because $V_1 = 0$, simplifies to

$$V_O = -V_2 \tag{7.2}$$

If we are interested in the voltage amplification (V_O/V_{IN}) provided by the circuit this property is conveniently expressed by substituting for V_2 in Equation (7.2) to obtain

$$V_O/V_{IN} = -(R_2/R_1) \tag{7.3}$$

So, by choosing suitable values for R_1 and R_2 (say, $R_1 = 1\,k\Omega$, $R_2 = 10\,k\Omega$) we can ensure that a ten times magnified version of V_{IN} appears at the output of the opamp, suggesting the value of the circuit as a voltage amplifier. In Figure 7.1 we have shown V_{IN} as a constant voltage, but in general it can vary with time, provided always that V_O is maintained within the saturation limits of the opamp.

7.2   The Inverter

In the foregoing analysis of the inverter circuit two assumptions have been made, and although they are often thought to be identical it is vital to realize that they are not. The first was to use the virtual short-circuit concept, ensuring that the voltage V^- is negligible. The second, *and entirely separate* assumption, is that no current flows into either of the opamp's two input terminals. We have made use of each of those two assumptions in the analysis of the inverter, and it is essential not to confuse or equate them. The assumption of zero input currents is equally valid for large-signal operation, whereas the virtual short-circuit concept is not.

The simple circuit of Figure 7.1 raises a number of questions. First, where does the current I flow after passing through R_2? Since nothing else is connected to the output of the opamp it must all flow into the opamp's output terminal. Second, what is the meaning of the minus sign in Equation 7.3? The minus sign does not imply that the circuit is not a good amplifier; it simply means that the sign of the output voltage is opposite to that of the input voltage V_{IN}. That is why the circuit is called an *inverter*. A third question concerns the contrast between the magnitude of the voltage amplification that the opamp can achieve (typically between 10^4 and 10^6) and the modest value of the inverter's gain (in our example, $|V_O/V_{IN}| = 10$). We seem to be throwing away a great opportunity! There are two answers to this third question.

Manufacturing variations

The first answer concerns the variability of manufactured artefacts. If you're manufacturing a large number of 'identical' items – whether they are hats or cups or electrical components such as resistors and opamps – the most noticeable and unwanted feature of what you produce is the fact that they are *not* identical! One hat will be a bit wider than another, even marginally; one cup will be a bit thinner than another, and the value of a resistor said to have a nominal value of $10\,k\Omega$ may well have an actual value somewhere between 9 and $11\,k\Omega$. In the case of an opamp the voltage amplification in the linear region may also vary quite widely from one opamp to the next. So, in order to design circuits having a reliable and predictable performance we must somehow minimize the effects of manufacturing variations.

This has in fact been achieved in the inverter circuit, because the voltage amplification of the opamp does not appear in the expression for V_O/V_{IN}. This does not mean that we can remove the opamp from the circuit and still achieve the same voltage amplification! It means that, to an extremely good approximation, variations in the opamp's gain will not affect the inverter's gain. That advantage is obtained at a price – the low gain of the circuit. But what about the inevitable variations in the value of the resistors? The voltage amplification V_O/V_{IN} will certainly be sensitive to their variation, except in one interesting – and very common – case. If the two resistors R_1 and R_2 are part of an integrated circuit, and if they are located close together on a chip, then they will tend to vary in the same way from one circuit to another, such that the ratio R_2/R_1, and hence the circuit's voltage amplification, remains essentially constant from one chip to another.

The second answer involves issues that we treat in Chapter 11, and will be discussed there.

A model

In any field it is often helpful to have a simple 'model' of something so that the prediction of how it will behave within an environment can be made easier. This is the case with the inverter, whose model is shown in Figure 7.4. We know from the analysis we have carried out above that the current flowing into the circuit from the voltage source V_{IN} is V_{IN}/R_1, and that is certainly the case in the model. We also showed that the output voltage V_O was a multiple $(-R_2/R_1)$ of the voltage V_{IN}: again, this is represented faithfully in the model by the voltage-controlled voltage source.

Stability

What happens if we accidentally make a mistake when connecting the opamp within the inverter circuit, so that R_2 is connected between the opamp output and the *positive* input terminal of the opamp, as shown in Figure 7.5? The analysis we have used to derive Equation 7.3 will not predict the outcome.

To explain qualitatively what happens we redraw the circuit of Figure 7.5 as shown in Figure 7.6. Purely for convenience we assume the two resistors have equal value, that the opamp has a voltage amplification in its linear region of 2000, and

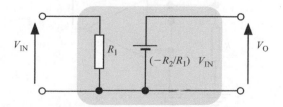

Figure 7.4 A model of the inverter circuit

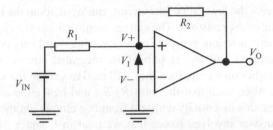

Figure 7.5   The inverter circuit with the opamp input terminals interchanged

that the source is temporarily removed. We now assume that, due to noise in the circuit or a signal picked up from a nearby radio, the voltage V_O increases by a very small amount – say, one microvolt. The voltage divider provided by the two resistors will then cause a voltage V^+ (and hence V_I) of 0.5 μV to appear at the positive input terminal of the opamp. If the opamp's voltage gain is 2000, this will cause the output voltage V_O to be 1000μV, or 1 mV. In turn, this voltage will be halved to provide V^+(0.5 mV) which will then be amplified to provide an output voltage of 1 V. This regenerative action continues very rapidly until V_O reaches and remains at its maximum value of +10 V and, by voltage divider action, V^+ is equal to 5 V. We refer to such behaviour as *instability*.

Restoration of the input voltage V_{IN} to the circuit of Figure 7.5 merely provides an initial starting point for the rapid change in V_O to one of its limiting values, V_S or $-V_S$.

Feedback

It will have been noticed in our discussion of opamp circuits so far that when the output voltage V_O is connected back to the positive input terminal of the opamp

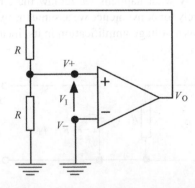

Figure 7.6   The circuit of Figure 7.5 redrawn

via a resistor, the circuit's state is liable to change rapidly, or be unstable, when certain conditions are satisfied. We have seen some examples in Chapter 6. We generally refer to this condition as *positive feedback*. By contrast the inverter circuit involves *negative feedback* (R_2 is connected between the output terminal and the negative input terminal) and is associated with a circuit that is stable. The theory of feedback, however, is quite complex and inappropriate for discussion here: the correlation of positive feedback with instability and of negative feedback with stability is mentioned here purely to aid in the correct identification of the expected behaviour of simple opamp circuits.

7.3 Noninverting Connection

Another useful circuit in which the opamp operates in its linear region is the noninverting connection and is shown in Figure 7.7: it is seen to involve negative feedback. Since no current is drawn by the negative input terminal we can use the voltage divider principle to express V^- in terms of V_O:

$$V^- = [R_1/(R_1 + R_2)]V_O \tag{7.4}$$

If the opamp is operating in its linear region we can assume a virtual short-circuit between the input terminals, so that $V^+ = V^-$. Because V_{IN} and V^+ are identical, we can write, from Equation (7.4), that

$$V_{IN} = V^+ = V^- = [R_1/(R_1 + R_2)]V_O$$

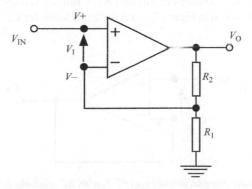

Figure 7.7 A noninverting circuit

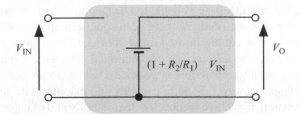

Figure 7.8   A model of the noninverting circuit of Figure 7.7

so that the voltage amplification of the circuit is

$$V_O/V_{IN} = 1 + R_2/R_1 \qquad (7.5)$$

Whatever values are chosen for R_1 and R_2 the output voltage V_O has the same sign as the input voltage V_{IN} (hence 'noninverting'). A model of the noninverting circuit is shown in Figure 7.8 and could be referred to as a voltage-controlled voltage source.

Voltage follower

Why are we interested in the noninverting connection? One reason is the fact that zero current is taken from the input voltage source V_{IN}. In other words, no 'load' is placed on that source. Another, and important reason, is that a special case of the circuit has useful properties, as we shall now show.

If we choose $R_2 = 0$ (a short-circuit) and $R_1 =$ infinity (an open-circuit) the circuit takes the form shown in Figure 7.9 and the expression for voltage amplification

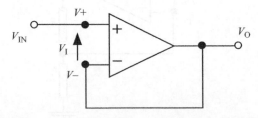

Figure 7.9   The noninverting circuit of Figure 7.7, with $R_1 =$ infinity and $R_2 = 0$, provides a voltage follower, with $V_O = V_{IN}$

(Equation 7.5) becomes

$$V_O/V_{IN} = 1 \qquad (7.6)$$

(in other words, the output voltage is the same as the input voltage).

At first this property of the circuit of Figure 7.9 seems singularly unattractive, since a single wire of zero resistance between input and output would achieve the same result at much lower cost! The reason why the circuit of Figure 7.9 is of interest lies not only in its voltage amplification, but also in the fact that it draws zero current from the input voltage V_{IN}, as an example will now show.

Imagine that we have a circuit represented by its Thevenin model – say a voltage V_X and an output resistance of $10\,k\Omega$ – as shown in Figure 7.10(a), and suppose we have the task of applying the same voltage V_X across a load resistance of $10\,k\Omega$. Making a direct connection, as in Figure 7.10(a), will have two important effects. First, only half the voltage V_X will appear across the load (and will vary if the load is varied in value). Second, a current will be drawn from the source containing V_X. Why does this matter? Because we may be trying to make measurements on a circuit without disturbing it, just as we do when placing the probe of an oscilloscope on a connection point within a circuit; drawing a current away from it will change its operation. The solution is shown in Figure 7.10(b); no current is drawn from the circuit being observed, so that $V_{IN} = V_X$. As we discovered in analysing the circuit of Figure 7.9, the voltage V_O is identical to V_{IN} and hence V_X, so that the voltage V_X appears directly across the load resistance, *whatever the value of that resistance*. We say that the opamp circuit is '*buffering*' the load from the circuit, and because $V_O = V_{IN}$ we often refer to the opamp circuit of Figure 7.9 as a *voltage follower*.

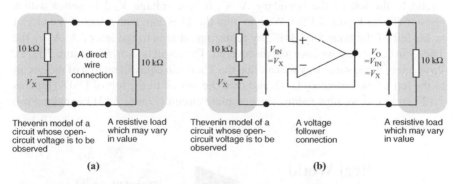

(a)   (b)

Figure 7.10   An example to show that although a direct connection (a) ensures identical voltages at each end of the *connection*, a voltage follower (b) ensures that no current is drawn from the circuit being observed and delivers a voltage V_X irrespective of the load resistance

7.4 Other Opamp Circuits Operating in the Linear Region

There are many circuits created by circuit designers to exploit the linear region of an opamp's characteristic. We examine three of them here: others will be encountered in the problems provided at the end of the chapter.

Digital-to-analog (D–A) conversion

Following digital processing it is often required to transform a quantity represented digitally into the single analog quantity that it represents (Figure 7.11). This process is called Digital-to-analogue (D–A) conversion. There are many circuits that can perform D–A conversion: the one we shall study makes use of the inverter circuit, and also illustrates how valuable the concept of a Thevenin equivalent circuit can be.

The circuit of a simple D–A converter is shown in Figure 7.12(a.) We assume that the digital representation of a quantity of interest is the sequence of voltages V_1 to V_4, each of which will take on one of two values representing binary 1 and 0. The voltage V_O at the output of the opamp is the analog equivalent. The value of R characterising the resistors of value R and $2R$ can have any value: it is the relationship between these resistors that matters.

At first sight the circuit appears complicated, but its operation is easily explained if we proceed to develop the Thevenin equivalent of the circuit to the left of the boundary D–D'.

To do this we begin by finding the Thevenin equivalent of the much simpler circuit to the left of the boundary A–A'. It is a voltage $V_4/2$ in series with a resistor R (see Figure 7.12b). We now find the Thevenin equivalent of the circuit to the left of the boundary B–B', by replacing what is to the left of A–A' by the Thevenin equivalent we have just discovered. The result is shown in Figure 7.12(c). Proceeding in the same way we can find the Thevenin equivalent, first of the circuit to the left of the boundary C–C' and then the circuit to the left of D–D' (Figures 7.12d and e). If we now redraw the complete circuit (Figure 7.13) by making use

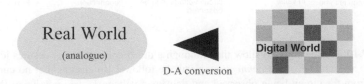

Figure 7.11   It is often necessary for information in digital form to be transformed to analog form

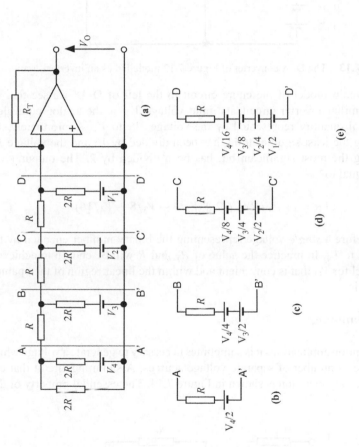

Figure 7.12 The circuit (a) of a D–A converter. Shown in (b), (c), (d) and (e) are the Thévenin models of that part of the circuit lying to the left of the various boundaries

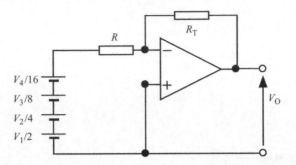

Figure 7.13 The D–A converter of Figure 7.12 modelled as an inverter circuit

of the Thevenin model of the entire circuit to the left of D–D' we see that we have the familiar inverter circuit: its input voltage V_{IN} is the analog equivalent of the digital quantity represented by the voltages V_1 to V_4. Appropriately, V_4, representing the least significant bit, has been divided by 16 and the voltage V_1, representing the most significant bit, has been divided by 2. The output V_O is therefore equal to

$$V_O = (-R_T/R)(V_1/2 + V_2/4 + V_3/8 + V_4/16) \tag{7.7}$$

and is therefore a *single* voltage representing the binary number encoded by the voltages V_1 to V_4. In practice the value of R_T and R will be chosen to achieve a voltage level for V_O that is convenient and within the linear region of the opamp's characteristic.

Voltage summing

In instrumention applications it is sometimes necessary to generate a voltage which is the sum of a number of separate voltage sources. A summing circuit that can perform such an operation is shown in Figure 7.14. The essential property of this

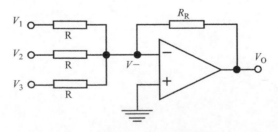

Figure 7.14 A summing amplifier

circuit derives from the virtual earth principle: since V^- can be assumed negligible the currents flowing into node X from the voltage sources are independent of each other and, together, flow through the resistor R_R. Thus,

$$V_O = -R_R(V_1/R + V_2/R + V_3/R)$$

A variant of the circuit of Figure 7.14, called a 'weighted summer', uses appropriate values of the resistors R so that the contribution of each voltage source to the output voltage V_O can be arranged.

Example 7.2

The circuit of Figure 7.15 is a weighted summer. By invoking the virtual earth concept we can find, by Ohm's Law, the total current from the three voltage sources flowing towards the negative input terminal of the opamp:

$$I = (V_1/2 + V_2/5 + V_3/10)\,\text{k}\Omega$$

All this current flows into the 1 kΩ resistor creating a voltage

$$V = I \times 1\,\text{k}\Omega = 0.5V_1 + 0.2V_2 + 0.1V_3$$

Since, by KVL, $V_O = -V$, the output voltage V_O is given by

$$V_O = -(0.5V_1 + 0.2V_2 + 0.1V_3)$$

showing that different weights have been attached to the various voltage sources. If the minus sign is unwelcome the voltage V_O can always be applied to an inverter designed to provide a voltage gain of -1.

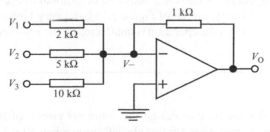

Figure 7.15 A weighted summer

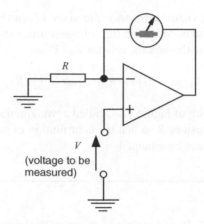

Figure 7.16   A high-input-resistance voltmeter

A voltmeter with high input resistance

We have already seen how the zero input current property of an opamp can be exploited. Another example is shown in Figure 7.16 which shows the circuit of a voltmeter having the very desirable property of a high input resistance. The meter measures current and will have an internal resistance: its value is immaterial, as we shall see.

The analysis of this circuit is straightforward, especially if, as a reminder of the virtual short-circuit phenomenon, we sketch a dashed line between the opamp's two input terminals. With a virtual short-circuit between the two opamp input terminals the voltage to be measured, V, is transferred to the negative input terminal of the opamp. Thus, the voltage across the resistor R is V, creating a current V/R. This current can only flow through the meter, and therefore the meter current is directly proportional to the voltage V. One advantage offered by this circuit is the fact that very little current flows into the positive input terminal of the opamp, thereby ensuring that the voltmeter circuit of Figure 7.16 has a very high input resistance. Another advantage is that the operation is unaffected by the resistance of the meter.

Example 7.3

Let us suppose that we have to design the voltmeter circuit of Figure 7.16. We shall assume that the meter has a full-scale deflection when 50 μA passes through it, and that we want that full-scale reading to indicate a value of 5 V for the voltage V.

To choose an appropriate value of R we note that when V has its maximum value of 5 V, the current through R is $5/R$. Since this current passes wholly through the meter, it should be equal to 50 μA. Thus, $5/R = 50$ μA, so that the required value of R is 100 kΩ.

7.5 Problems

Linear operation

Problem 7.1

For each of the circuits shown in Figure P7.1 find the value of the voltage V. Assume that the limits to the output voltage of the opamp are $+10$ and -10 V.

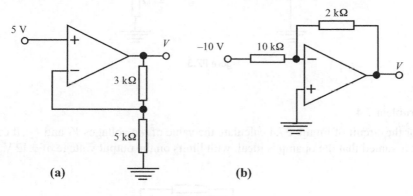

(a)   (b)

Figure P7.1

Problem 7.2

For the circuit shown in Figure P7.2 calculate the value of the voltage V. It can be assumed that the opamp is ideal, with output voltage limits of ± 12 V.

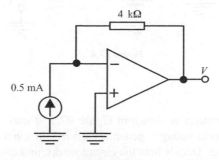

Figure P7.2

Problem 7.3

A circuit is shown in Figure P7.3. First, remove any components that are redundant. Then calculate the value of the voltage V. The limits to the opamp's output voltage are ± 8 V.

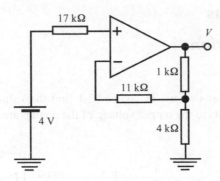

Figure P7.3

Problem 7.4

For the circuit of Figure P7.4 calculate the value of the voltages V_1 and V_2. It can be assumed that the opamp is ideal, with limits on the output voltage of ± 12 V.

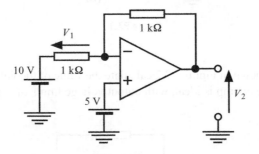

Figure P7.4

Problem 7.5

The circuit of a lightmeter is shown in Figure P7.5. It uses a photodiode which, with a constant reverse voltage, generates $0.5\,\mu$A of current per microwatt of incident radiant power. Decide how the circuit works, and choose a value for R so that the scale factor on the voltmeter is $2.5\,\mu$W/mV.

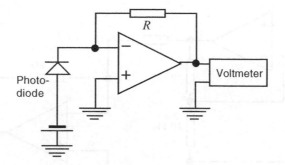

Figure P7.5

Problem 7.6

For the circuit of Figure P7.6 find expressions for (in this order) V_A, V_B, V_C and V_{OUT}.

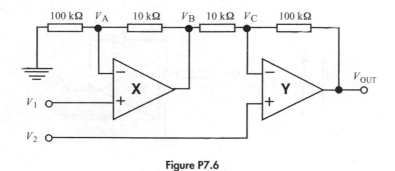

Figure P7.6

Amplifiers using opamps

Problem 7.7

The circuit of a popular instrumentation amplifier is shown in Figure P7.7. Find, in terms of V_1 and V_2: (a) the current through the $10\,k\Omega$ resistor; (b) the voltages V_X and V_Y; and (c) the voltage V_O. It can be assumed that all opamps are working in the linear region of operation.

Problem 7.8

It can be useful to generate a voltage which is the logarithm of another voltage. That is the function of the logarithmic amplifier circuit shown in Figure P7.8. The

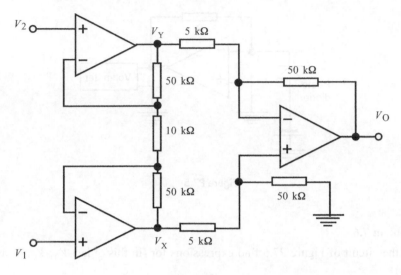

Figure P7.7

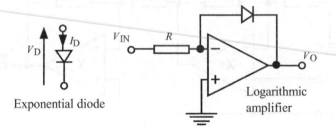

Exponential diode

Logarithmic amplifier

Figure P7.8

voltage–current ($I_D \sim V_D$) relation of the so-called exponential diode is:

$$I_D = I_S[\exp(V_D/V_T) - 1]$$

where I_S is known as the reverse saturation current of the diode and V_T (=25 mV at room temperature) is the 'thermal voltage'.

Show that if V_{IN}/R is much greater than the reverse saturation current I_S of the diode, and if the opamp is operating in its linear region, then

$$V_O = -V_T \ln(V_{IN}/RI_S)$$

Problem 7.9

Find the Thevenin equivalent circuit of the circuit within the shaded area in Figure P7.9.

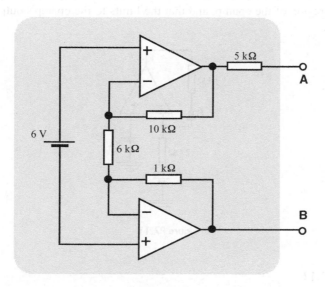

Figure P7.9

Problem 7.10

Use the superposition principle to express the voltage V_O in the circuit of Figure P7.10 in terms of the two voltages designated V_1 and V_2.

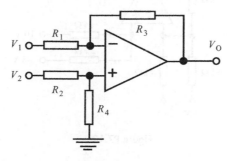

Figure P7.10

The circuit must be designed to act as a difference amplifier: in other words, V_O must be proportional to $V_2 - V_1$. Derive the relation between the resistances R_1, R_2, R_3 and R_4 for this result to be achieved.

Problem 7.11

For the circuit of Figure P7.11 choose one value of the voltage V that will ensure that the current I lies between the limits of 1.5 and 2.5 mA. Assume operation in the linear region of the opamp, and that the limits to the opamp's output voltage are ± 10 V.

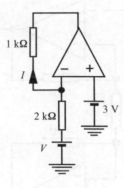

Figure P7.11

Problem 7.12

Determine the value of V in the circuit of Figure P7.12. Assume that the limits to the opamp's output voltage are ± 10 V.

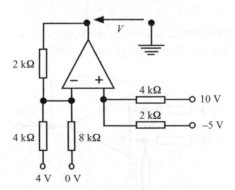

Figure P7.12

Mixed and Dynamic Opamp Circuits

In the two previous chapters we have seen the range of useful behaviour that can be achieved through the inclusion of one or more opamps in a circuit, in some cases exploiting the linear region of an opamp's characteristic and in others the existence of two well-defined states. It will not be surprising that *combinations* of these two types of circuit can yield additional useful application possibilities. But a component that can be combined with both types of opamp circuit to yield even more useful types of behaviour is the *capacitor*. Before proceeding, therefore, we examine the characteristics of the capacitor.

8.1 The Capacitor

The symbol for the two-terminal device called a capacitor is shown in Figure 8.1. We use v and i to denote the voltage and current of a capacitor, conventionally using the reference directions shown. In many useful circuits the voltage v and the current i will not be constant as in a DC circuit, and we indicate that fact by using lowercase letters to denote possibly time-varying quantities. Often we shall explicitly emphasize the time-varying nature of the capacitor voltage and current by writing $v(t)$ and $i(t)$, respectively.

For an ideal capacitor the voltage $v(t)$ and current $i(t)$ are related by the equation

$$i(t) = C\,dv(t)/dt \qquad (8.1)$$

Introductory Circuits   Robert Spence
© 2008 John Wiley & Sons, Ltd

Figure 8.1　The symbol representing a capacitor. The constant *C* is its capacitance in farads

The single constant *C* that characterizes a given capacitor is called its *capacitance*. The units are Farads, named after Michael Faraday.

Equation (8.1) tells us that (Figure 8.2) if a constant current *I* is applied to a capacitor, the capacitor voltage will increase linearly with time, at a rate determined by the current *I* and the capacitance *C*. If we examine a more realistic situation where the capacitor voltage does not increase indefinitely (Figure 8.3) we see what happens when the current into the capacitor is zero. Equation (8.1) tells us that if *i(t)* is zero, the rate of change of capacitor voltage is also zero; in other words, the capacitor voltage remains constant. This implies that if the current source were to be removed (equivalent to setting its value to zero), the voltage on an ideal capacitor will remain – for ever – at the last value achieved until more current is supplied. We then say that the capacitor has been 'charged' to a particular value: with an ideal capacitor it retains that charge. When more current is supplied, as at time T in Figure 8.3, the capacitor voltage will start moving away from its existing value, again according to Equation (8.1).

One more comment about the implication of Equation (8.1) is required before we begin to exploit the unique properties of a capacitor. It has to do with sudden changes in voltage like those experienced when a Schmitt trigger changes state (see Figure 6.10). Equation (8.1) tells us that an instantaneous change in capacitor voltage requires an infinite current. When, as in practice, such a current is unavailable, the capacitor voltage cannot change instantaneously. Thus if, in a circuit (Figure 8.4), a voltage at one terminal A of a capacitor changes instantaneously,

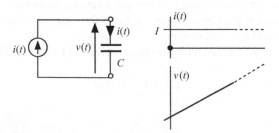

Figure 8.2　The application of a constant current to a capacitor, and the resulting voltage response

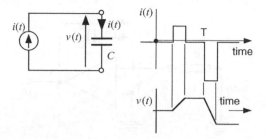

Figure 8.3 Response of a capacitor to a time-varying current source

the voltage at the other terminal B will change by the same amount, consistent with the capacitor voltage remaining constant.

The unique electrical nature of a capacitor, encapsulated in Equation (8.1) and illustrated in Figures 8.2 to 8.4, enables a circuit designer to create some very useful circuits.

8.2 The Integrator

The circuit of Figure 8.5 is very similar to the inverter circuit encountered in the previous chapter except that we have replaced resistor R_2 with a capacitor. What useful function can this new circuit perform?

Let us assume that the opamp is operating in its linear region (we shall see soon that this is the case), so that the negative input terminal of the opamp is a virtual earth, i.e., $v^- = 0$. From Ohm's law we see that the current $i(t)$ through R_1 is

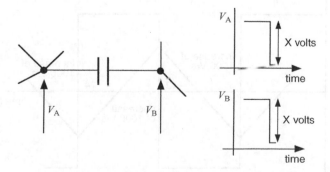

Figure 8.4 If the voltage at one terminal of a capacitor changes instantaneously, the voltage at the other terminal will exhibit the same change since, instantaneously, a capacitor voltage does not change

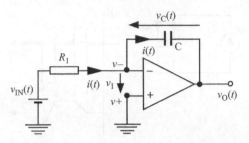

Figure 8.5   An integrating circuit

$v_{IN}(t)/R_1$. For the moment we shall assume that V_{IN} is constant. Since no current can enter the negative input terminal of the opamp, all of this current flows into the capacitor, as indicated in Figure 8.5. According to Equation (8.1) the voltage v_C across the capacitor increases linearly at the rate $(v_{IN}/R_1)/C$. Since our principal interest is in the output voltage v_O we apply KVL in the same way that we did in Figure 7.2 to find that $v_O = -v_C$ (because $v_I = 0$). The output voltage v_O therefore *decreases* linearly, at the rate of $-(v_{IN}/R_1)/C$.

If the voltage v_{IN} in the circuit of Figure 8.5 is at first constant, as illustrated in Figure 8.6, v_O will continue to decrease at a constant rate. Now let us suppose that instantaneously, at time $t = 0$ as shown in Figure 8.6, v_{IN} changes sign, but not magnitude. The current into the capacitor will now have the same magnitude, but the opposite sign, so v_O will now *increase* at the rate $(v_{IN}/R_1)/C$. Further regular changes in v_{IN} will thereby result in a triangular waveform for v_O.

We have described the circuit of Figure 8.5 by working out its detailed behaviour when driven by the particular waveform shown in Figure 8.6. But there is a simpler

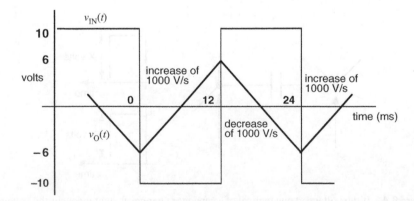

Figure 8.6   The input and output voltages of the integrator of Figure 8.5 (numerical values refer to Example 8.1)

description of the circuit: it is an *integrator*. Equation (8.1) shows that current (directly proportional to the input voltage v_{IN} in the circuit of Figure 8.5) is the derivative of the capacitor voltage (which apart from a sign is the output voltage v_O). The output voltage v_O is, therefore, apart from the sign, the integral of the input voltage, multiplied by a constant determined by the resistor and capacitor in the circuit.

Example 8.1

We consider the integrator circuit of Figure 8.5, with $R_1 = 10\,\mathrm{k\Omega}$ and $C = 1\,\mu\mathrm{F}$. The input waveform v_{IN} is defined by the numerical values shown in Figure 8.6. We are required to find the waveform of the output voltage v_O.

From Ohm's law we know that, while $v_{IN} = 10\,\mathrm{V}$, $i(t) = 10\,\mathrm{V}/10\,\mathrm{k\Omega} = 1\,\mathrm{mA}$. Recalling that $v_O = -v_C$ we can write, from Equation (8.1), that

$$\mathrm{d}v_O(t)/\mathrm{d}t = -i(t)/C = -10^{-3}/10^{-6} = -1000\,\mathrm{V/s}.$$

The output voltage will fall until the time (designated as $t = 0$ in Figure 8.6) at which v_{IN} changes sign. The voltage v_O therefore begins to increase at the rate $1000\,\mathrm{V/s}$ until, at $t = 12\,\mathrm{ms}$, v_{IN} again changes sign: since v_{IN} is now $10\,\mathrm{V}$, the voltage v_O will again decrease at the rate of $1000\,\mathrm{V/s}$. For the given waveform of v_{IN} the waveform of v_O will be as shown in Figure 8.6.

8.3 Dynamic Opamp Circuits

An illustration of the new possibilities opened up by the use of a capacitor is provided by the circuit of Figure 8.7. At first this circuit appears to be a very confusing collection of components, but insight into its operation is easily gained if it is recognized as the combination of two functional blocks, as suggested by the grey shading. We immediately recognize an integrator of the type just discussed and the alternative trigger circuit introduced in Chapter 6. The output of the integrator (v) provides the input to the trigger, and the trigger's output (v_O) in turn provides the input to the integrator. To explain the operation of the circuit we assume the component values shown in Figure 8.7, and that the limits to the output voltage of each opamp are $+10$ and $-10\,\mathrm{V}$.

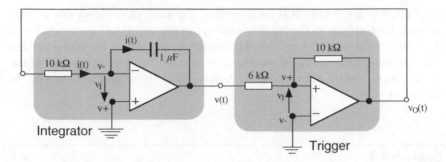

Figure 8.7   The circuit analysed in Example 8.2

Example 8.2

The output of the trigger in Figure 8.7 can only be at $+10$ or -10 V. To begin our analysis we shall assume that $v_O = 10$ V, so that the input voltage to the integrator has the same value. If the integrator opamp is operating in the linear region we can apply the concept of the virtual earth and assume that $v^- = 0$. Using Ohm's law we find the current $i(t)$:

$$i(t) = (10 - 0)/10\,\mathrm{k\Omega} = 1\,\mathrm{mA}$$

From our earlier discussion we know that the rate of change of the output voltage of the integrator, here denoted as $v(t)$, will be

$$\mathrm{d}v(t)/\mathrm{d}t = -10^{-3}/10^{-6} = -1000\,\mathrm{V/s}.$$

The voltage $v(t)$ is also the input to the trigger. We know that, at some point, the input to the trigger will be sufficiently negative to cause its output to change from $+10$ to -10 V. To determine this threshold value for $v(t)$ we refer to a calculation (Equation 6.1) made for the same circuit in Chapter 6: it is -6 V. The decreasing output voltage $v(t)$ of the integrator will eventually reach this value, causing the output voltage $v_O(t)$ of the trigger to change from $+10$ to -10 V, as shown in Figure 8.8.

The input voltage to the integrator has now changed sign, but has the same magnitude, so its output voltage $v(t)$ now begins to *increase* at 1000 V/s. However, in view of the changed state of the trigger, the threshold voltage that $v(t)$ needs to achieve to cause a change of state back to an output voltage of $+10$ V has now changed to $+6$ V. The cycle then repeats (Figure 8.8). Only one thing remains to be done, and that is to check whether, for the integrator's opamp, our assumption of operation in its linear region is valid. Inspection of the waveform of $v(t)$ in Figure 8.8 shows that it is.

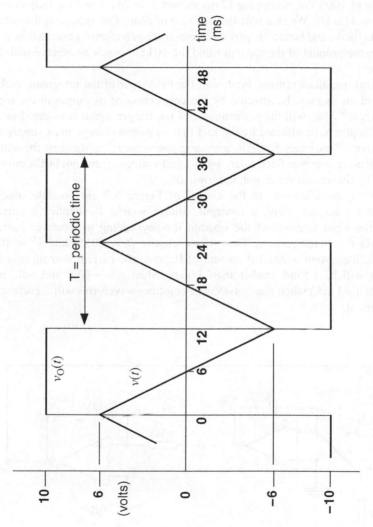

Figure 8.8   The voltages v_O and v in the circuit of Figure 8.7

From the resulting waveforms of v_O and v shown in Figure 8.8 we see that the circuit generates both square and triangular voltage waveforms, both of which find use in a wide variety of applications. To calculate the frequency of these waveforms we observe that half the periodic time T involves a change of 12 V at the rate of 1000 V/s, occupying 12 ms so that $T = 24$ ms and the frequency is $1000/24 = 41.6$ Hz. We also note that the rate of change of voltage at the output of the integrator, and hence the periodic time of the waveforms generated, is controlled by the product of the capacitor and the $10 \text{ k}\Omega$ resistor associated with the integrator.

Two final questions remain. First, will the behaviour of the integrator, earlier considered on its own, be affected by the connection of its output to the input of the trigger? Also, will the performance of the trigger, again considered on its own in Chapter 6, be affected by the fact that its output voltage must supply the input current of the integrator? The answer in each case is 'no' because the output voltage of an opamp is defined solely by its input voltage v_1, and not by the current supplied by the opamp via its output terminal.

Interesting modifications to the circuit of Figure 8.7 are possible: one is shown in Figure 8.9. Here, a constant voltage source V supplies a current V/R to the input terminal of the opamp, thereby adding to whatever current arrives via the $10 \text{ k}\Omega$ resistor. Thus, if we choose $R = 20 \text{ k}\Omega$ and $V = 10 \text{ V}$ to provide a constant additional current of 0.5 mA, the current flowing into the capacitor will be 1.5 mA (rather than 1 mA) when $v = +10 \text{ V}$ and -0.5 mA (rather than -1 mA) when $v = -10 \text{ V}$. The resulting waveforms will therefore be asymmetrical.

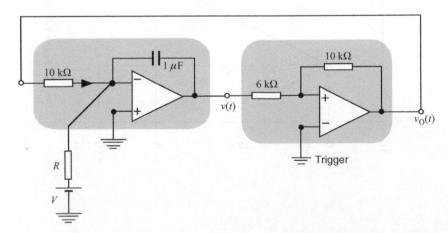

Figure 8.9   Modification to the circuit of Figure 8.7, resulting in asymmetrical waveforms

8.4 Problems

The capacitor

Problem 8.1

For the circuit of Figure P8.1(a) the current source $i(t)$ has the waveform shown in Figure P8.1(b). Sketch, on the same plot, the waveform of the capacitor voltage $v(t)$ whose value is zero at time $t = 0$.

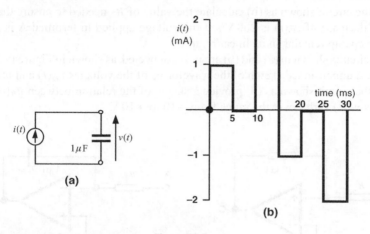

(a)

(b)

Figure P8.1

Problem 8.2

For the circuit of Figure P8.1(a) the voltage waveform shown in Figure P8.2 is observed. Deduce the waveform of the current source.

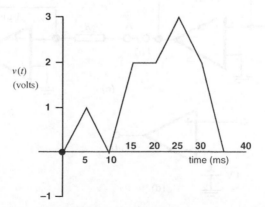

Figure P8.2

Dynamic circuits

Problem 8.3

The operational amplifiers shown in Figure P8.3 can all be assumed ideal, with their output voltages saturating at ± 10 V.

For the circuit shown as (a) choose a value for R_1 to ensure that the voltage v changes between -10 and $+10$ V when the voltage $v_{IN}(t)$ falls below -8 V or rises above $+8$ V.

For the circuit shown as (b) calculate the value of R_2 needed to ensure that the rate of decrease of $v_O(t)$ is 200 V/s if the voltage applied to terminal A is 10 V and the opamp remains in its linear region.

The circuits shown in (a) and (b) are now connected, as shown in Figure P8.3(c). Provide a dimensioned sketch of the waveforms of the voltages $v_O(t)$ and $v(t)$.

For the circuit shown as (d) provide a sketch of the relation between $v_B(t)$ and $v_O(t)$ as $v_O(t)$ varies over the range from -10 to $+10$ V.

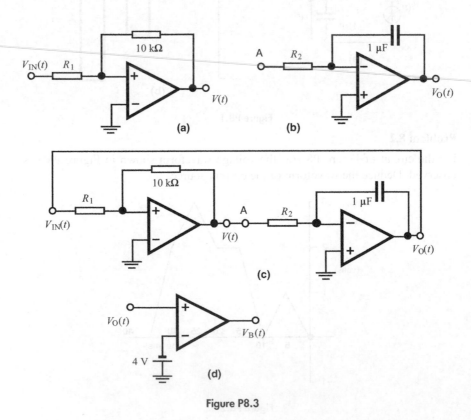

Figure P8.3

The circuit shown in (d) is now connected to the circuit of (c) at the point whose voltage is denoted $v_O(t)$ in both circuits. Provide a dimensioned sketch of the resulting waveform of $v_B(t)$; it may be convenient to show the waveform of $v_B(t)$ on the previously sketched plot of the waveforms of circuit (c).

Problem 8.4

Design a circuit using opamps, resistors and a capacitor that will generate a square-wave voltage having a peak-to-peak amplitude of 20 V and a periodic time of 10 ms. The opamps available are characterized by limits of ± 10 V on the output voltage. Your answer should include the circuit diagram with component values, and should identify the point at which a square-wave voltage appears.

Problem 8.5

Refer to Figure P8.3. In part (c) the negative input terminal of the integrator's opamp is now additionally connected to a resistor of 100 kΩ and a DC voltage of 10 V, as shown in Figure P8.5. Calculate the new periodic time of the voltage waveform appearing at A and provide a sketch of the voltage waveforms $v_O(t)$ and $v(t)$.

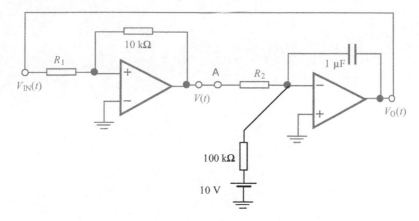

Figure P8.5

Overview: AC Circuits

There are many extremely useful circuits comprising, or modelled by, a combination of resistors, *capacitors* and *inductors*: they are usually referred to as *AC circuits* (AC = 'alternating current'). Capacitors and inductors are called reactive components, and their presence in a circuit means that we cannot use, directly, the methods of analysis we developed for DC circuits. Also, whereas the sources in DC circuits are constant, circuits containing capacitors and inductors are usually driven by sources that are not constant: frequently they are sinusoidal in nature. Typical examples include the filters that separate the different speech channels in a telecommunication system.

The main problem we face is how best to discuss these circuits and predict their behaviour. We could, for example, draw graphs of all the sinusoidal voltages and currents in a circuit, or perform a trigonometric analysis, but these approaches are tedious and error-prone and don't provide much insight.

A much better idea by far is to represent each sinusoidal voltage and current by a single complex number, because the marvellous consequence of so doing is that we can use all the techniques (KCL, KVL, superposition, etc.) that we learned in the context of DC circuits. The only difference is that we have to handle complex quantities rather than real ones.

A related method of representing sinusoidal voltages and currents – the *phasor diagram* – also helps us to visualize AC circuit behaviour.

By making use of complex analysis we can easily compute the way in which the AC performance of a circuit varies with the frequency of any voltage or current sources; an example is the way in which the amplification of your HiFi varies from bass to treble. This is often referred to as *frequency domain* performance.

Introductory Circuits   Robert Spence
© 2008 John Wiley & Sons, Ltd

9

AC Circuits and Phasor Diagrams

Previous chapters have considered circuits in which voltages and currents are constant or, in the case of Chapters 6 and 8, move between well-defined constant values. There is, however, enormous interest in the way that circuits behave when voltages and currents are *sinusoidal* in form.

Why is this? There are two reasons. For a long time the source of electrical power for many applications has been a sinusoidally varying voltage, typically with a frequency of 50 or 60 Hz. The other reason is that many of the signals involved in communication systems can be considered to be the addition of a number of sinusoids. For example (Figure 9.1) a square wave can be approximated by the addition of four sinusoids of appropriate frequency and amplitude. Anyone purchasing a HiFi amplifier, for example, will be concerned with its ability to amplify the highest note from a piccolo as well as the lowest note from a 64 foot organ pipe.

In circuits designed to perform appropriately when the voltage and current sources are sinusoidal rather than constant, two components – called *reactive components* – are of special interest. We have already met one of them, the *capacitor*, in Chapter 8. The other is the *inductor*, which we introduce below.

9.1 Reactive Components

The capacitor

We encountered our first reactive component in the form of a capacitor in Chapter 8, where it was found to be useful in circuits in view of its ability to integrate. Our

Introductory Circuits   Robert Spence
© 2008 John Wiley & Sons, Ltd

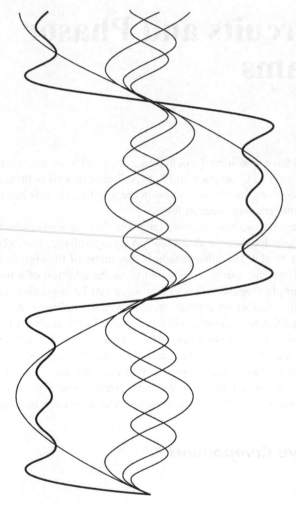

Figure 9.1 The addition of four sinewaves of relative amplitudes 1, 1/3, 1/5 and 1/7, and relative frequencies ω, 3ω, 5ω and 7ω (light-lines) yields an approximation (bold line) to a square waveform.

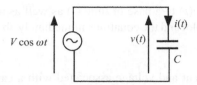

Figure 9.2 A capacitor with its voltage defined by a sinusoidal voltage source of radian frequency ω

interest now is in its behaviour when the voltage across it is sinusoidal. To begin to investigate why this is so we connect one to a sinusoidally varying voltage $V \cos \omega t$, as shown in Figure 9.2. This figure also introduces the symbol for an independent voltage source of sinusoidal waveform. In the expression for the sinusoidal voltage source:

$$v(t) = V \cos \omega t \tag{9.1}$$

V is known as the *amplitude* of the sinusoidal voltage source, and ω the radian frequency.

Equation (8.1) describing a capacitor can be rewritten as

$$i_C(t) = C dv_C(t)/dt \tag{9.2}$$

showing us that, for this simple circuit,

$$i(t) = C dv_C(t)/dt = Cd(V \cos \omega t)/dt = -\omega C V \sin \omega t \tag{9.3}$$

Equation (9.3) can be rewritten as

$$i(t) = \omega C V \cos(\omega t + \pi/2)$$

or

$$i(t) = I \cos(\omega t + \pi/2)$$

where

$$I = \omega C V \tag{9.4}$$

and is the *amplitude* of the sinusoidal current.

If we plot $i(t)$ and $v(t)$ to a base of time (t) as well as angle (ωt) we arrive at Figure 9.3. From this plot and the equations immediately above we can make some useful observations:

- The sinusoidal current and voltage associated with a capacitor have the same frequency;
- For the capacitor Equation (9.4) describes a component relation which has the same form as Ohm's law;
- There is a phase relation between the current and voltage: the capacitor current *leads* the voltage in the sense that, for example, it reaches its maximum before the voltage does.

The first observation, along with the same property we shall derive for an inductor, means that if a circuit is connected to a sinusoidal source having a frequency ω, then all voltages and currents in the circuit have the same frequency. This property enormously simplifies the analysis of a circuit.

The second observation is also very important. Having studied the analysis of DC circuits, it is encouraging to find a component relation having the same form as Ohm's law. If we know the voltage amplitude we only need to divide by a constant (albeit one which is a function of frequency) to find the current amplitude. We shall see the advantage of this relation soon.

Although the third observation is easily derived from the waveforms of Figure 9.3 we shall show that there is a graphical technique that makes it much easier to visualize the phase relations among a number of waveforms.

Example 9.1

In the circuit of Figure 9.4 a sinusoidal voltage source whose amplitude is 4 V and whose frequency is 159 Hz is applied to a 1 μF capacitor. Find an expression for the capacitor current.

The frequency 159 Hz is, to a very good approximation, a radian frequency of 1000. From the relation between current and voltage for a capacitor we can write

$$i(t) = C\,dv(t)/dt = 10^{-6} \times d(4\cos 1000t)/dt = -10^{-6} \times 4$$
$$\times 1000 \sin 1000t = -4 \times 10^{-3} \sin 1000t \text{ A}$$
$$= -4 \sin 1000t \text{ mA}.$$

In other words, the capacitor current has an amplitude of 4 mA and a frequency of 1000 rad/s ($=159$ Hz). Note that the amplitude of a sinusoidal source is often

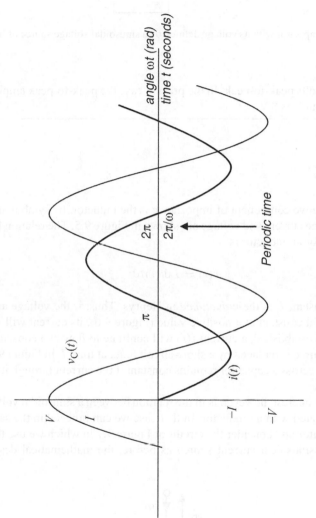

Figure 9.3 The waveforms of sinusoidal voltage across, and sinusoidal current through, a capacitor

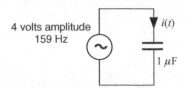

Figure 9.4   A capacitor with its voltage defined by a sinusoidal voltage source of frequency 159 Hz

quoted as N volts peak-to-peak. In the present case the peak-to-peak amplitude of the voltage source is 8 V.

The inductor

The other reactive component of importance is the inductor. Its symbol and associated reference current and voltage are shown in Figure 9.5. The relation between $v(t)$ and $i(t)$ for an inductor is

$$v(t) = L\,\mathrm{d}i(t)/\mathrm{d}t \tag{9.5}$$

where the constant L is the *inductance* in Henrys. Thus, if the voltage across an inductor is held constant at a positive value (Figure 9.6), its current will increase linearly. Once established, a current $i(t)$ will continue to flow at a constant rate if the voltage source is replaced by a short-circuit (e.g., at time T in Figure 9.6), just as the voltage across a capacitor remains constant if the current through it is set to zero.

Our current interest, however, is in the relation between a sinusoidal voltage and current associated with an inductor. In that case we can proceed in the same way as for a capacitor and consider the circuit of Figure 9.7 in which we use the same symbol for a sinusoidal current source as before, the mathematical description

Figure 9.5   An inductor

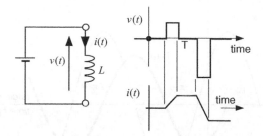

Figure 9.6 Reponse of an inductor to a time-varying voltage source

making it clear that it has a sinusoidal waveform. To show that an arbitrary phase angle θ makes no difference to the outcome, we assume that the current source is described by

$$i(t) = I \cos(\omega t + \theta) \tag{9.6}$$

Recalling Equation (9.5) we can write:

$$v(t) = L \, di(t)/dt = \omega L I[-\sin(\omega t + \theta)] = \omega L I \cos[\omega t + \pi/2 + \theta] \tag{9.7}$$

so that $v(t) = V \cos(\omega t + \pi/2 + \theta)$
where

$$V = \omega L I \tag{9.8}$$

If we plot $i(t)$ and $v(t)$ to a base of time (t) as well as angle (ωt) we arrive at Figure 9.8. From this plot and from the equations immediately above we can make some useful observations:

• The sinusoidal current and voltage associated with an inductor have the same frequency;

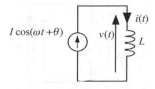

Figure 9.7 An inductor with its current defined by a sinusoidal current source of radian frequency ω

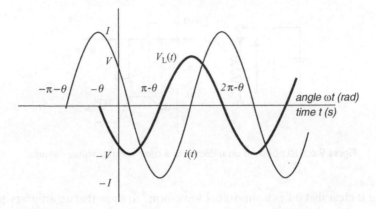

Figure 9.8 The waveforms of sinusoidal current through, and sinusoidal voltage across, an inductor

- For an inductor the component relation $V = \omega L I$ has the same form as Ohm's law;
- There is a phase relation between the current and voltage: the inductor current *lags* the voltage in the sense that, for example, it reaches its maximum after the voltage does.

The resistor

The resistor is *not* a reactive component: nevertheless it is essential that we be clear about its behaviour when its voltage and current are sinusoidal. If, as in Figure 9.9, we apply a sinusoidal voltage source $V \cos \omega t$ to a resistor, the current $i(t)$ will, by Ohm's law, be given by

$$i(t) = (V/R) \cos \omega t$$

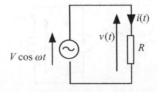

Figure 9.9 A resistor with its voltage defined by a sinusoidal voltage source of radian frequency ω

which we can write as

$$i(t) = I \cos \omega t$$

where

$$I = V/R \qquad (9.9)$$

As for the capacitor and inductor, we can make three observations about the sinusoidal current and voltage associated with a resistor: that a sinusoidal current results from the application of a sinusoidal voltage of the same frequency; that Ohm's law relates the amplitudes of current and voltage; and that there is no phase difference between the current and voltage waveforms.

Summary

Before proceeding further it is useful to summarize what we know about the components we shall call R (resistance), L (inductance) and C (capacitance) when their currents and voltages are sinusoidal:

- All currents and voltages have the same frequency as the sinusoidal source
- All component relations have the same form as Ohm's law
- There are specific phase differences between current and voltage for L and C

Simple circuits

What we have learned so far allows us to analyse simple circuits containing more than one component, simply by employing the fundamental relations describing the capacitor and inductor. An example will provide an illustration.

Example 9.2

Figure 9.10 shows a circuit containing a resistor and an inductor in series with a sinusoidal current source. It is required to find an expression for the voltages across the resistor and inductor and hence across the current source.

The resistor voltage is found by the application of Ohm's law, and is

$$v_R(t) = 100 \times 2 \cos (1000t + 20°) = 200 \cos (1000t + 20°) \, \text{mV}$$

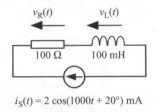

$i_S(t) = 2 \cos(1000t + 20°)$ mA

Figure 9.10   The circuit discussed in Example 9.2

The inductor voltage is found from Equation (9.3) to be

$$v_L(t) = L\,di(t)/dt = (100 \times 10^{-3})\,2000 \cos(1000t + 110°)$$
$$= 200 \cos(1000t + 110°)\,\text{mV}.$$

Noting that these two voltages have identical amplitudes and differ in phase by 90° we can express their sum, the voltage $v_S(t)$ across the current source, as

$$v_S(t) = 282.8 \cos(1000t + 65°)\,\text{mV}$$

9.2   The Phasor Diagram

The question now arises as to how to determine the behaviour of a circuit containing a number of reactive components and a source of sinusoidal voltage or current. At the same time we look for a representation of sinusoidal currents and voltages that can help one to visualize, perhaps better than may be possible with mathematical equations, what is happening in such a circuit.

Fortunately, there is a representation of sinusoidal voltages and currents that can provide useful insight into their relative magnitudes and phases. It is called the *phasor diagram*. What is more, the phasor concept it embodies provides a good stepping stone to the next chapter in which a very powerful method of analysis is presented.

We start by considering the simple circuit of Figure 9.11: a capacitor of capacitance C whose voltage is determined by a sinusoidal voltage source $v(t)$ of amplitude V and radian frequency ω:

$$v(t) = V \cos \omega t$$

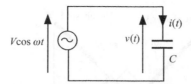

Figure 9.11   A capacitor with its voltage defined by a sinusoidal voltage source of radian frequency ω

In the previous chapter we established that the resulting current $i(t)$ is

$$i(t) = I \cos{(\omega t + \pi/2)}$$

where I, the amplitude of the current, is given by

$$I = \omega C V$$

The *phasor diagram* representation of $v(t)$ and $i(t)$ is shown in Figure 9.12. The two phasors labelled **V** and **I** have lengths proportional to the amplitudes V and I of $v(t)$ and $i(t)$ respectively, the scale of proportionality being chosen to provide a conveniently sized diagram. The phasors rotate anticlockwise at an angular frequency ω. It is their *projection* onto a stationary reference axis that identifies the *actual* values of $v(t)$ and $i(t)$, as shown in Figure 9.12. Because the phasors are rotating at the frequency ω, the projections also vary at the frequency ω, with an amplitude determined by the lengths of the phasors. Because the phasors are rotating, what we have in Figure 9.12 is a 'snapshot' taken at a particular instant of time.

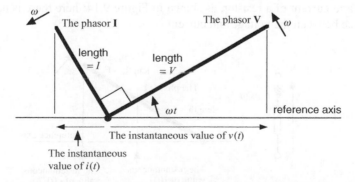

Figure 9.12   A phasor diagram showing phasors representing capacitor voltage and current. Note that the phasors rotate at an angular frequency ω and it is their projection onto a reference axis that determines the actual scalar instantaneous values of current and voltage

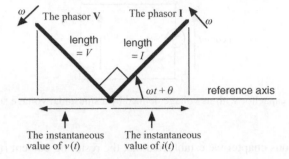

Figure 9.13   A Phasor diagram showing phasors representing inductor voltage and current

The phasor diagram of Figure 9.12 places clearly in evidence the relative phase of the current and voltage, and the fact that the capacitor current leads the voltage by 90°. In this sense it does not matter how the reference axis is oriented, because projections onto it will still have the same amplitudes and relative phase; the only effect is to change the time origin, and since the sinusoidal voltage and current are continuous this is immaterial.

The phasor representation for the voltage and current of an inductor (see Equations 9.6 and 9.7):

$$i(t) = I \, \cos(\omega t + \theta)$$
$$v(t) = \omega L I \, \cos(\omega t + \theta + \pi/2)$$

is shown in Figure 9.13 from which it is clear that an inductor current *lags* the voltage by 90°.

We must not forget that a phasor diagram can also represent the sinusoidal voltage and current of a resistor, as shown in Figure 9.14: here there is no phase difference between the voltage and current.

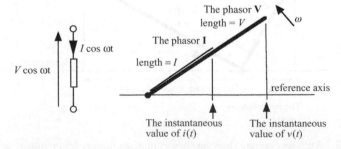

Figure 9.14   The phasors representing the sinusoidal voltage and current of a resistor are in phase. The current phasor has been offset slightly for clarity

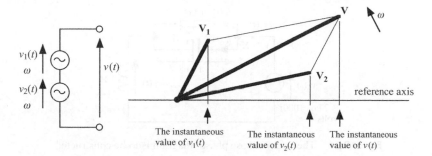

Figure 9.15   A Phasor diagram showing that the phasor addition of voltages obeys KVL. Hairlines are used to indicate construction

Kirchhoff's laws

Figures 9.12, 9.13 and 9.14 have provided us with representations of component relations. But what happens when we connect those components together? For DC circuits the answer was provided by Kirchhoff's laws; fortunately, they do so again for sinusoidal voltages and currents.

That Kirchhoff's voltage law applies to phasors can be illustrated by the simple example of Figure 9.15 in which two sinusoidal voltage sources of radian frequency ω are connected in series. We wish to represent, by a phasor, the resulting voltage which, according to KVL, is the sum of the two source voltages. Each of the two rotating phasors V_1 and V_2, when projected onto the reference axis, describes one of the two sinusoidal voltages $v_1(t)$ and $v_2(t)$. Their *phasor* (also known as *vector*) addition, shown as the phasor V, is also seen to provide a projection onto the reference axis which defines the scalar time-varying value of the voltage $v(t)$. Figure 9.15 thereby illustrates the fact that KVL applies to voltage phasors. In a similar way it can be shown that KCL also applies to current phasors. What Figure 9.15 also shows by example is that voltage *amplitudes* do **not** in general obey KVL: the same applies to current amplitudes.

9.3   Constructing a Phasor Diagram

We can now combine knowledge of *component* properties, as represented by phasors, and the fact that KCL and KVL apply to *connections*, to describe a *circuit* by means of a phasor diagram. To illustrate how this is done we consider the circuit shown in Figure 9.16.

To draw the phasor diagram we start (Figure 9.17) with the voltage across the capacitor (Why? The answer is given later in this section). We arbitrarily draw

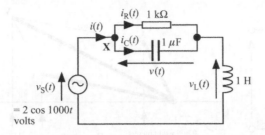

Figure 9.16   The circuit whose phasor diagram is to be constructed

(Figure 9.17a) its phasor **V** along the reference line. We do not yet know its amplitude so we simply indicate that the length of this phasor is V.

The voltage $v(t)$ appears across the resistor so the phasor $\mathbf{I_R}$ representing the current through it is parallel to **V** (i.e., it is 'in phase' with it). Again we have to choose what we hope is a suitable length (Figure 9.17b) for the phasor $\mathbf{I_R}$. However, we do know the relation between the amplitude of the current and voltage of the resistor ($I = V/R$), so we can label the length of the phasor $\mathbf{I_R}$ as $V/1000$.

The voltage $v(t)$ also appears across the capacitor. Knowing that capacitor current leads voltage we know the direction in which to draw (Figure 9.17c) the current phasor $\mathbf{I_C}$ and we can also compute its length from Equation (9.4):

$$I = \omega C V = 10^3 \times 10^{-6} V = V/1000$$

so that the phasors $\mathbf{I_R}$ and $\mathbf{I_C}$ are of equal length.

We now invoke KCL at node X in the circuit of Figure 9.16. The phasor $\mathbf{I}$ representing the current $i(t)$ is (Figure 9.17d) the phasor addition of $\mathbf{I_R}$ and $\mathbf{I_C}$ and therefore has a length $V\sqrt{2}/1000$ and subtends an angle of $45°$ with the phasor **V**.

The current $i(t)$ also flows through the inductor, so the phasor $\mathbf{V_L}$ representing (Figure 9.17e) the voltage $v_L(t)$ across the inductor is $90°$ in advance of the phasor **I**. Its length can be found from the equation 9.8 ($V = \omega L I$) to be $1000 \times 1 \times V\sqrt{2}/1000$ which is $V\sqrt{2}$.

In view of the dimensions of the phasors **V** and $\mathbf{V_L}$ it is easy to see that their phasor addition (Figure 9.17f) to obtain the phasor $\mathbf{V_S}$ by KVL shows that $\mathbf{V_S}$ is $90°$ in advance of **V** and has the same magnitude.

It is at this point that we can refer to the actual amplitude (2 V) of the voltage source $v_S(t)$ and thereby calculate the amplitude (2 V) and phase ($90°$ lagging with respect to $v_s(t)$) of the voltage $v(t)$ represented by the phasor **V**. In the same way, for example, we can calculate the amplitude of the current through the resistor as $V/1000 = 2/1000 = 2\,\text{mA}$. In other words, the current $i_R(t)$ in the resistor can

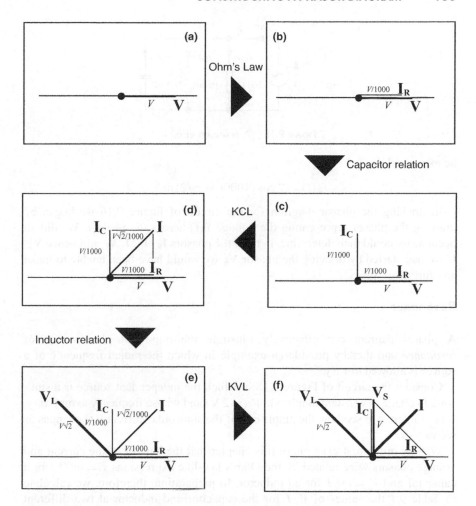

Figure 9.17 Steps in the construction of a phasor diagram for the circuit of Figure 9.16. Where possible phasor lengths have been indicated beside the phasor. Some overlapping phasors have been offset slightly for clarity. Hairlines are included to indicate constructions

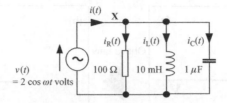

Figure 9.18   A resonant circuit

be expressed as

$$i_R(t) = 2\cos(1000t - \pi/2)\,\text{mA}$$

In drawing the phasor diagram for the circuit of Figure 9.16 we began by drawing the phasor representing the voltage $v(t)$ across the resistor. We did so because we could then determine, in turn, the phasors $\mathbf{I}_R$, $\mathbf{I}_C$, $\mathbf{I}$, $\mathbf{V}_L$ and hence $\mathbf{V}_S$. If we had started by drawing the phasor $\mathbf{V}_S$ we would have been unable to make any further progress.

Resonance

A phasor dagram can effectively illustrate the important phenomenon of *resonance* and thereby provide an example in which the radian frequency of a source is allowed to vary.

Consider the circuit of Figure 9.18 in which the independent source is a sinusoidal voltage of constant amplitude V (= 2 V) and whose frequency ω may vary. It is of interest to see how the amplitude of the sinusoidal current $i(t)$ changes as we vary ω.

We saw in the first example of this chapter that the lengths of the current and voltage phasors were related by the Ohm's law-like expressions $I = \omega C V$ for a capacitor and $V = \omega L I$ for an inductor. In preparation, therefore, we calculate in Table 9.1 the values of V/I for the capacitor and inductor at two different frequencies.

Table 9.1   Relevant to the analysis of the circuit of Figure 9.18

ω radians/sec	ωC siemens	$1/\omega C$ ohms	ωL ohms
10^4	10^{-2}	100	100
$10^4\sqrt{2}$	$\sqrt{2}/100$	$100/\sqrt{2}$	$100\sqrt{2}$

The values of V/I for the inductor and capacitor of Figure 9.18 for two values of ω, the radian frequency of the current source

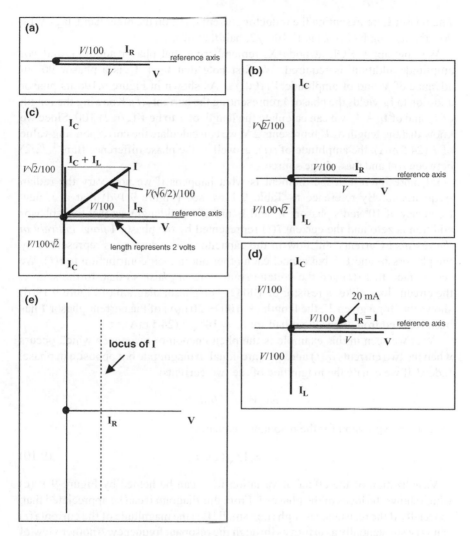

Figure 9.19 Construction of phasor diagrams for the circuit of Figure 9.18. Some over-lapping phasors have been offset for clarity

To investigate the circuit of Figure 9.18 we first assume that $\omega = 10^4 \sqrt{2}$ rad/s. We start the phasor diagram (Figure 9.19a) by arbitrarily drawing the phasor **V**, of length V, representing $v(t)$ along a reference axis. The phasor $\mathbf{I_R}$, representing the current $i_R(t)$ in the resistor of 100 Ω is then in phase with **V** as shown and its length is $V/100$.

At the radian frequency $\omega = 10^4 \sqrt{2}$, the phasor $\mathbf{I_C}$ representing $i_C(t)$ leads **V** by 90° (Figure 9.19b) and its length will be ωCV which for $\omega = 10^4 \sqrt{2}$ is $V\sqrt{2}/100$.

The phasor $\mathbf{I_L}$ representing the inductor current $i_L(t)$, on the other hand, lags $\mathbf{V}$ by $90°$ and its length $(V/\omega L)$ is $V/100\sqrt{2}$, half that of $\mathbf{I_C}$.

We now apply KCL at node X, remembering that phasor addition, and *not* amplitude addition, is required. We first note that $\mathbf{I_C} + \mathbf{I_L}$ is a phasor $90°$ in advance of $\mathbf{V}$ and of amplitude $V/100\sqrt{2}$. As shown in Figure 9.19c its phasor addition to $\mathbf{I_R}$ yields the phasor $\mathbf{I}$ representing the current $i(t)$. Knowing the length of $\mathbf{I_R}$ and of $\mathbf{I_C} + \mathbf{I_L}$ we can calculate the length of I to be $V(\sqrt{6/2})/100$. Since we know that the length of V represents 2 V, we can calculate the corresponding value of I (24.5 mA), the amplitude of $i(t)$, as well as the phase difference $(\tan^{-1}\sqrt{2/2})$ between $i(t)$ and the voltage source.

Of more interest at the moment is what happens if we now vary the radian frequency ω. By reference to Table 9.1 we see (Figure 9.19d) that at a new frequency of 10^4 rad/s phasors $\mathbf{I_C}$ and $\mathbf{I_L}$ are of equal length, so that their phasor addition is zero and the current $i(t)$ represented by the phasor $\mathbf{I}$ *flows entirely in the resistor*. Currents still flow in the capacitor and inductor (as represented by the phasors $\mathbf{I_C}$ and $\mathbf{I_L}$), but cancel each other out in their contribution to $i(t)$. We see, in fact, that $i(t)$ and the voltage source are *in phase* so that, to the source, the circuit 'looks like' a resistor of 100 Ω. The phasor diagram of Figure 9.19(d) shows that for a given V, the length $(V/100 = 20\,\text{mA})$ of the current phasor $\mathbf{I}$ has been reduced from the value it had for $\omega = 10^4\sqrt{2}$ (24.5 mA).

What we seen in this example is the phenomenon of *resonance* which occurs when the two currents $i_C(t)$ and $i_L(t)$ are equal in magnitude but opposite in phase. Indeed, if we equate the magnitude of the two currents:

$$\omega C V = V/\omega L$$

we find an expression for the resonant frequency:

$$\omega = 1/\sqrt{(LC)} \tag{9.10}$$

Visualization of the effect of variation in ω can be helped by Figure 9.19(e) which shows the locus of the phasor $\mathbf{I}$. From this diagram it can be appreciated that, especially if the resistance is high (i.e., small $|\mathbf{I_R}|$) the magnitude of the current $i(t)$ can vary substantially as ω moves through the resonant frequency. Another view of the resonance phenomenon can be obtained if we examine a circuit identical to that of Figure 9.18 but with the sinusoidal voltage source replaced by a current source of amplitude 1 mA. Resonance is again experienced, but now with the voltage $v(t)$ exhibiting the variation with frequency sketched in Figure 9.20. At resonance the capacitor and inductor currents 'cancel each other out' and the voltage is due entirely to the current $1\cos\omega t$ mA flowing in the 100 Ω resistor. It will be noted that, for convenience, logarithmic scales are adopted for amplitude and frequency, a technique introduced and fully explained in Chapter 11.

The examples presented above are important for two reasons. First, to show how phasor diagrams can help us to visualize circuit behaviour over a range

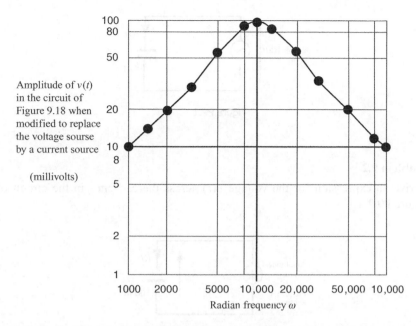

Figure 9.20 Pertinent to the circuit of Figure 9.18 in which the sinusoidal voltage source is replaced by a sinusoidal current source of 1 mA amplitude. The sketch shows the variation of the ampitude of the voltage $v(t)$ as the source frequency ω varies

of frequencies. Second, through the phenomenon of resonance, to introduce the important concept of *selectivity*. If, in place of the single current source to which Figure 9.20 refers, we had a collection of current sources in parallel representing a number of TV transmissions, then only that transmission occurring within a specified 'bandwidth' determined by the values of R, L and C would generate a significant voltage amplitude: in other words, all other frequencies lying outside that bandwidth would to a greater or lesser degree be *filtered* out. In telephone communication systems, where we need to separate different conversations, the principle of filtering is again employed, though usually with circuits containing many inductors and capacitors.

9.4 Problems

Single capacitors and inductors

Problem 9.1

Derive an expression for the current $i(t)$ flowing in the capacitor in the circuit of Figure P9.1. Sketch at least one cycle of the voltage and current waveforms, indicating values of both time and angle (ωt) on the horizontal axis.

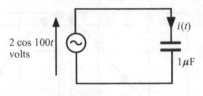

Figure P9.1

Problem 9.2

Derive an expression for the voltage $v(t)$ across the capacitor in the circuit of Figure P9.2

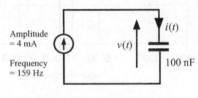

Figure P9.2

Problem 9.3

Derive an expression for the voltage $v(t)$ across the inductor in the circuit of Figure P9.3.

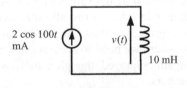

Figure P9.3

Problem 9.4

As shown in Figure P9.4, a resistor and a capacitor are connected in parallel across the terminals of a sinusoidal voltage source. Derive expressions for the currents in the resistor and capacitor, and hence the total current supplied by the voltage source.

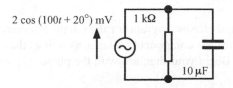

Figure P9.4

Phasor diagrams

Problem 9.5

Sketch a dimensioned phasor diagram representing all the currents, as well as the voltage, in the circuit of Figure P9.5.

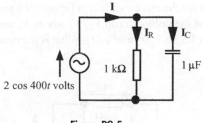

Figure P9.5

Problem 9.6

Sketch a dimensioned phasor diagram showing all component currents and voltages in the circuit of Figure P9.6.

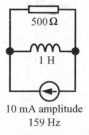

Figure P9.6

Problem 9.7

Sketch a dimensioned phasor diagram for the circuit of Figure P9.7, and thereby obtain expressions for the component voltages as well as the current supplied by the voltage source. Begin your diagram with the phasor representing the resistor voltage.

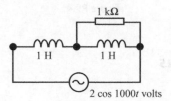

Figure P9.7

Problem 9.8

Draw a phasor diagram for the circuit shown in Figure P9.8 and hence determine the magnitude and phase of the voltage ratio V_2/V_1 where V_1 and V_2 are, respectively, the amplitudes of the voltages $v_1(t)$ and $v_2(t)$. What is the phase difference between $v_1(t)$ and $v_2(t)$?

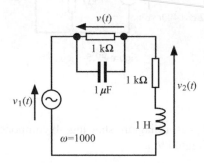

Figure P9.8

10

Complex Currents
and Voltages

While the concept of phasor diagrams introduced in Chapter 9 provides a useful representation of sinusoidal currents and voltages that helps one to visualize circuit behaviour, it has some disadvantages:

- Phasor diagrams require *graphical construction*, unsuited to computers;

- They refer to a *single frequency* of operation, notwithstanding examples (such as the resonant circuit treated in the previous chapter) where performance over a frequency range can be visualized to some extent;

- The 'starting phasor' in a phasor diagram construction can be difficult to identify or may not even exist.

There is, therefore, a need for a much simpler way of analysing circuits in which voltages and currents are sinusoidal, one which is amenable to computer implementation and one that can easily handle performance over a range of frequencies.

10.1 Euler's Theorem

The basis of the principal approach to AC circuit analysis is Euler's theorem, which states that:

$$e^{j\theta} = \cos\theta + j\sin\theta \tag{10.1}$$

Introductory Circuits   Robert Spence
© 2008 John Wiley & Sons, Ltd

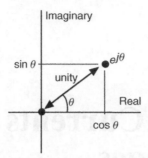

Figure 10.1   Representation of $e^{j\theta}$ in the complex rane

Here, j rather than i is used to denote the square root of minus one to avoid confusion with the symbol for current. The representation of $e^{j\theta}$ in the complex plane with real and imaginary axes is shown in Figure 10.1. We see that the real component of $e^{j\theta}$ is a cosine term which can conveniently be used to describe a sinusoidal voltage or current having a value such as 4 cos ωt. So if we choose $\theta = \omega t$, the real part of $e^{j\theta}$ is cos ωt, a term that has often appeared in our expressions for current and voltage.

With $\theta = \omega t$, Euler's theorem (Equation 10.1) becomes:

$$e^{j\omega t} = \cos \omega t + j \sin \omega t \tag{10.2}$$

an expression whose graphical representation is shown in Figure 10.2. Note that the point $e^{j\omega t}$ is not stationary; like the phasors we have met it rotates in a circular trajectory of unity radius at a radian frequency ω. Therefore, whereas we earlier described a sinusoidal voltage by an expression such as

$$v(t) = V \cos(\omega t + \theta)$$

we can now write, by reference to Figure 10.2,

$$v(t) = \text{Re}[Ve^{j(\omega t + \theta)}] = \text{Re}[Ve^{j\theta}e^{j\omega t}]$$

The term $Ve^{j\theta}$ is a constant and in general complex: we shall call it **V**, so that we can write

$$v(t) = \text{Re}[\mathbf{V}e^{j\omega t}] \tag{10.3}$$

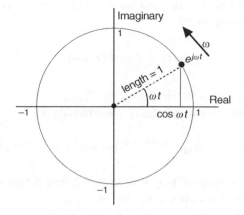

Figure 10.2   Properties of $e^{j\omega t}$ in the complex plane

where we use the bold **V** to denote what we shall call the *complex voltage*. The close relation between the phasor representation of a voltage (by a rotating line) and the complex representation by a point describing a circular trajectory is clear, and justifies use of the same symbol (**V**) in both cases.

The obvious question at this point is 'why do we need to involve complex quantities? Why can't we just use trigonometric functions and write, for example, $v(t) = 4\sin 100t$ V?'. The answer lies in the fact that, as we shall shortly see, complex voltages and currents obey similar laws to the ones that govern voltages and currents in a DC circuit, with the valuable bonus that we can use the same approach to AC circuit analysis as the one we learned in Chapters 3 and 4 for DC analysis!

10.2   Component Relations

The capacitor

As presented in Chapter 8, the fundamental relation between the current and voltage of a capacitor is

$$i(t) = C\,\mathrm{d}v(t)/\mathrm{d}t$$

where C is called the capacitance and has the units of farads. Thus, if the voltage across a capacitor is expressed as

$$v(t) = V\cos(\omega t + \theta)$$

then $i(t) = \omega CV\cos(\omega t + \theta + \pi/2)$.
Using Euler's relation we can express $i(t)$ alternatively as

$$i(t) = \omega CV\,\mathrm{Re}[e^{j(\omega t + \theta + \pi/2)}] = \mathrm{Re}[\omega CVe^{j(\theta + \pi/2)}e^{j\omega t}]$$

In Chapter 9 we denoted ωCV by I. If we now denote the constant $Ie^{j(\theta + \pi/2)}$ by $\mathbf{I}$ (a complex quantity), we can express $i(t)$ as

$$i(t) = \mathrm{Re}[\mathbf{I}e^{j\omega t}] \tag{10.4}$$

in order to achieve the same form as the expression for voltage (Equation 10.3):

$$v(t) = \mathrm{Re}[\mathbf{V}e^{j\omega t}] \tag{10.3}$$

If we are given the value of $\mathbf{I}$ or $\mathbf{V}$ as well as the radian frequency ω we can easily determine, from Equation (10.4) or (10.3), the actual value of $i(t)$ or $v(t)$.

In a DC circuit we found that Ohm's law defined the relation between the current and voltage of a resistor. It is helpful now to discover the relation between the complex voltage $\mathbf{V}$ and the complex current $\mathbf{I}$ imposed by a capacitor.

We defined

$$\mathbf{I} = \omega CVe^{j(\theta + \pi/2)} = \omega CVe^{j\theta}e^{j\pi/2} \tag{10.5}$$

and we also defined

$$\mathbf{V} = Ve^{j\theta} \tag{10.6}$$

Combining these two relations we find

$$\mathbf{I} = \omega CVe^{j\pi/2}$$

But since, from Euler's theorem, $e^{j\pi/2} = j$, we can now write

$$\mathbf{I} = (j\omega C)\mathbf{V} \tag{10.7}$$

The importance of this relation is that it has the same form as Ohm's law. In exactly the same way we can show that the relation between the complex current **I** of an inductor and the complex inductor voltage **V** (both defined as in Equations 10.3 and 10.4) is

$$\mathbf{V} = (j\omega L)\mathbf{I} \tag{10.8}$$

Also, for a resistor,

$$\mathbf{V} = R\mathbf{I} \tag{10.9}$$

The last three equations show that, just as Ohm's law relates DC currents and voltages, a relation of the same form exists between the complex currents and voltages of capacitors, inductors and resistors. There is therefore considerable motivation now to see if complex currents obey Kirchhoff's current law and complex voltages obey Kirchhoff's voltage law. If that is the case, the same approach developed for DC circuit analysis can be used for AC circuit analysis.

10.3 Interconnection

We consider the case of three sinusoidal currents flowing into a node (Figure 10.3). According to KCL,

$$i_1(t) + i_2(t) + i_3(t) = 0$$

If all currents are sinusoidal and have the same frequency we can rewrite the expression as

$$I_1 \cos(\omega t + \phi_1) + I_2 \cos(\omega t + \psi_2) + I_3 \cos(\omega t + \phi_3) = 0$$

By reference to Euler's theorem we can rewrite this equation as

$$\mathrm{Re}[I_1 e^{j\phi_1} e^{j\omega t}] + \mathrm{Re}[I_2 e^{j\phi_2} e^{j\omega t}] + \mathrm{Re}[I_3 e^{j\phi_3} e^{j\omega t}] = 0$$

or, with reference to Equation (10.4),

$$\mathrm{Re}[\mathbf{I}_1 e^{j\omega t}] + \mathrm{Re}[\mathbf{I}_2 e^{j\omega t}] + \mathrm{Re}[\mathbf{I}_3 e^{j\omega t}] = 0$$

Figure 10.3   Relevant to the discussion of whether KCL is valid for complex currents

where $\mathbf{I}_1$, $\mathbf{I}_2$ and $\mathbf{I}_3$ are the complex currents representing the actual sinusoidal currents $i_1(t)$, $i_2(t)$ and $i_3(t)$. The last equation can be rewritten as

$$\text{Re}[(\mathbf{I}_1 + \mathbf{I}_2 + \mathbf{I}_3)e^{j\omega t}] = 0$$

Since $e^{j\omega t}$ is never equal to zero, it follows that

$$\mathbf{I}_1 + \mathbf{I}_2 + \mathbf{I}_3 = 0$$

In other words, for the example considered, Kirchhoff's current law holds for complex currents. While KCL holds for complex currents $\mathbf{I}_x$ and instantaneous currents $i_x(t)$, it does not hold for current amplitudes I_x.

In a similar way it can be illustrated by example that Kirchhoff's voltage law holds for complex voltages.

10.4   AC Circuit Analysis

The important conclusion we can now draw is that, for a circuit in which all currents and voltages are sinusoidal of a given frequency, the constraints upon voltages and separately upon currents imposed by connection, and the relations between the currents and voltages imposed by individual components, are of *precisely* the same form as for DC circuits if the sinusoidal currents and voltages are represented by their complex values $\mathbf{I}$ and $\mathbf{V}$, as expressed in Equations (10.3) and (10.4). It follows, therefore, that analysis can proceed in the same way as for DC circuits. The similarities between DC and AC currents and voltages are set out in Table 10.1.

To illustrate AC circuit analysis, and to emphasize that it has much in common with DC analysis, we shall work through a simple example in Example 10.1 below.

Table 10.1 Similarities between the component relations and connection constraints in DC circuits and AC circuits

DC Currents and voltages			AC Currents and voltages	
representation	*relation*		*relation*	*representation*
		Connections		
	$\sum_{node} I = 0$	*at node*	$\sum_{node} \mathbf{I} = 0$	
	$\sum_{loop} V = 0$	*around loop*	$\sum_{loop} \mathbf{V} = 0$	
		Components		
	$V = R\,I$	*resistor*	$\mathbf{V} = R\,\mathbf{I}$	
		capacitor	$\mathbf{I} = j\omega C\,\mathbf{V}$	
		inductor	$\mathbf{V} = j\omega L\,\mathbf{I}$	
	$V = constant$	*voltage source*	$\mathbf{V} = constant$	
	$I = constant$	*current source*	$\mathbf{I} = constant$	

Example 10.1

Our task is to determine the waveforms of the voltages and currents in the circuit shown in Figure 10.4a

In view of what we have discovered about the ease with which complex currents and voltages can be predicted, we now redraw the circuit and show complex currents and voltages (Figure 10.4b). For convenience we label two nodes as A and B.

For comparison we carry out, simultaneously, the DC analysis of a circuit having the same layout as the AC circuit to be analysed (Figure 10.4c).

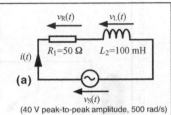

(a)

$v_S(t)$
(40 V peak-to-peak amplitude, 500 rad/s)

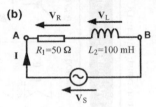

Figure 10.4 Relevant to a demonstration of the similarities between DC and AC analysis

In the centre we indicate the steps taken. A new term, explained later, is introduced in **boldface**. Quotes are used to draw attention to similarities between equations.

DC analysis	centre	AC analysis
$V_R = R_1 I$	Ohm's Law	$V_R = R_1 I$
$V_L = R_2 I$	Ohm's Law 'Ohm's Law'	$V_L = (j\omega L_2) I$
$V_S = V_R + V_L$	KVL	$V_S = V_R + V_L$
$R_{AB} = R_1 + R_2$	Equivalence (1)	$Z_{AB} = R_1 + j\omega L_2$
Total resistance between A and B		Total **impedance** between A and B
$V_S = 20$ volts		$V_S = 20 + j0$ volts
$R_1 = 50$ ohms	Values	$R_1 = 50$ ohms
$R_2 = 50$ ohms		$\omega L_2 = 50$ ohms
$R_{AB} = 100$ ohms	Substitution in 1	$Z_{AB} = 50 + j50$ ohms
$I = V_S/R_{AB} = 20/100 = 0.2$ amps	Ohm's Law 'Ohm's Law'	$I = V_S/Z_{AB} = (20 + j0)/(50 + j50)$ $= 0.2 - j0.2$ amps
$V_R = R_1 I = 50 \times 0.2 = 10$ volts	Ohm's Law 'Ohm's Law'	$V_R = R_1 I = 50 \times (0.2 - j0.2)$ $= 10 - j10$ volts
$V_L = R_2 I = 50 \times 0.2 = 10$ volts	Ohm's Law 'Ohm's Law'	$V_L = j\omega L_2 I = j50 \times (0.2 - j0.2)$ $= 10 + j10$ volts
$V_R + V_L = 20 = V_S$	Check	$V_R + V_L = 20 + j0 = V_S$

DC analysis

At this point the DC analysis is complete

AC analysis

At this point the AC analysis is *in*complete because sinusoidal currents and voltages are represented by complex quantities. We have reaped the benefit of such a representation (simple analysis, just like DC analysis) but now we must convert the complex currents and voltages to the sinusoidal currents and voltages they represent.

We recall that we have represented a current i(t) in the form

$$i(t) = \text{Re}[\mathbf{I}e^{j\omega t}]$$

Consider the current flowing through the resistor and inductor. Its complex representation is

$$\mathbf{I} = 0.2 - j0.2 \text{ A}$$

which can be represented by a point in the complex plane, as shown in Figure 10.5(a).

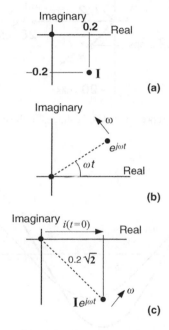

Figure 10.5 Steps in the transformation from a complex current to the actual value at a particular time

To find $i(t)$, however, we need the product of $\mathbf{I}$ and $e^{j\omega t}$ (see Equation 10.4), the latter quantity described by the point shown in Figure 10.5(b). Recalling that the product of two complex numbers is obtained by multiplying their magnitudes and adding their angles, and for convenience selecting $t = 0$, we obtain the representation of Figure 10.5(c).

We now have the actual value of $i(t)$ at time $t = 0$: it is the projection of the point $\mathbf{I}e^{j\omega t}$ on the real axis. More generally we see that the current $i(t)$ is a sinusoid having an amplitude of $0.2\sqrt{2}$ V. It is useful also to show all other voltages and currents in the complex plane (Figure 10.6) from which we can see, for example, that the amplitude of $v_{\text{L}}(t)$ is 14.14 V and that it leads the source voltage $v_{\text{S}}(t)$ by 45°. If required, waveforms can easily be derived from the diagram of Figure 10.6

as we have shown for $v_S(t)$ and $i(t)$ in Figure 10.7. At this point the AC analysis of the circuit of Figure 10.4(a) can be said to be complete.

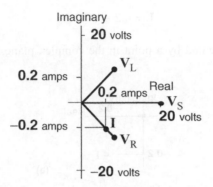

Figure 10.6   Complex representation of voltages in the circuit of Figure 10.4a

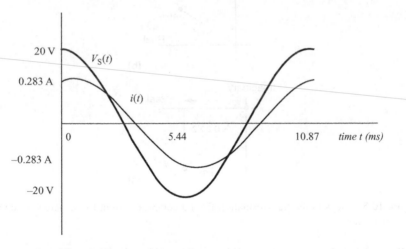

Figure 10.7   One cycle of the waveforms of $v_S(t)$ and $i(t)$

Having carried out the analysis and compared it to the analysis of a DC circuit of similar form we can make a number of useful observations and also clarify some points.

10.5 Observations

First, by standing back from detail we can summarize our approach to AC analysis by the diagram of Figure 10.8. The original problem is formulated in the *time*

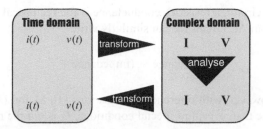

Figure 10.8 Illustrating the approach to AC analysis based on the use of complex currents and voltages

domain: we identify currents and voltages which are functions of time and we wish to describe them in detail. To do that we *transform* our problem into the *complex domain* in which we can easily calculate the complex representations of all currents and voltages. Finally we transform our results back into the time domain.

Next, we used the term *impedance* to describe the relation between a complex current and a complex voltage. Just as in DC circuits where we called the ratio V/I the resistance, in the complex domain we are also interested in the ratio $\mathbf{V}/\mathbf{I}$ and we use the term impedance. Since impedance is a complex quantity we need to talk about its real and imaginary parts which we call, respectively, *resistance* and *reactance*. Thus:

$$\text{impedance} = \text{resistance} + j(\text{reactance}) \tag{10.7}$$

and has the dimensions of ohms. Typically we denote the quantities in Equation (10.7) by $\mathbf{Z}$, R and X as in

$$\mathbf{Z} = R + jX \tag{10.8}$$

Sometimes, when dealing with DC circuits, we referred to the conductance of a resistor when discussing the ratio I/V: similarly, when dealing with AC circuits, we often wish to discuss the ratio $\mathbf{I}/\mathbf{V}$ in which case we define that ratio as:

$$\mathbf{I}/\mathbf{V} = \text{admittance} = \text{conductance} + j(\text{susceptance}) \tag{10.9}$$

often written as:

$$\mathbf{Y} = G + jB \tag{10.10}$$

Just as, in a DC circuit, a resistor's conductance is the reciprocal of its resistance ($G = R^{-1}$), for an AC circuit we can similarly write

$$\text{admittance} = (\text{impedance})^{-1} \tag{10.11}$$

or $\mathbf{Y} = \mathbf{Z}^{-1}$. However, with reference to Equations (10.8) and (10.10), we must be careful to note that, except in special conditions, G is *not* the reciprocal of R.

Example 10.2

Find the impedance and admittance between the terminals A and B for each of the two circuits shown in Figure 10.9 at a frequency of 1000 rad/s.

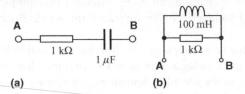

(a)   (b)

Figure 10.9   Relevant to Example 10.2

In the circuit of Figure 10.9(a) there are two components connected in series. To find the total impedance we add the impedances of the two components just as we would add the resistances of two resistors connected in series. The impedance of the resistor has only a real part, equal to $1 \text{k}\Omega$. The impedance of the capacitor is imaginary, and equal to $1/j\omega C$ (see Table 10.1). For the given values of ω and C the impedance of the capacitor is $-jk\Omega$. The impedance between A and B (we shall call it $\mathbf{Z}_{AB}$) is the sum of these two impedances and is therefore given by:

$$\mathbf{Z}_{AB} = 1 - j1 \,\text{k}\Omega$$

The admittance $\mathbf{Y}_{AB}$ between terminals A and B is the reciprocal of $\mathbf{Z}_{AB}$ and is therefore given by

$$\mathbf{Y}_{AB} = 1/\mathbf{Z}_{AB} = 1/(1 - j1) = 0.5 + j0.5 \,\text{mS}$$

In the circuit of Figure 10.9b there are two components connected in parallel. So we find the total admittance between A and B by proceeding as we would with two conductances connected in parallel: that is, by adding the admittance of the resistor (1 mS) to the admittance of the inductor ($1/j\omega L$) to get

$$\mathbf{Y}_{AB} = 1 - j0.1 \text{ mS}$$

Then, to obtain the impedance between A and B we find the reciprocal of $\mathbf{Y}_{AB}$:

$$\mathbf{Z}_{AB} = 1/(1 - j0.1) = (1 + j0.1)/1.01 \text{ k}\Omega = 0.99 + j0.99 \text{ k}\Omega$$

Example 10.3

Figure 10.10 shows the circuit for which we derived a phasor diagram in the previous chapter. We shall now predict its behaviour using the complex approach to AC analysis.

We immediately represent all currents and voltages by complex quantities (Figure 10.11). We also calculate, for the radian frequency of the source, the reactance of the inductor ($\omega L = 1 \text{ k}\Omega$) and the susceptance of the capacitor ($1/\omega C = 10^{-3}$ S).

First we calculate the impedance between X and Y (recall the expression for the resistance of two resistors in parallel):

$$\mathbf{Z}_{XY} = \frac{R(1/j\omega C)}{R + (1/j\omega C)} = 0.5 - j0.5 \text{ k}\Omega$$

Alternatively, we could first have calculated the admittance $\mathbf{Y}_{XY}$ between X and Y (recalling the equation for the conductance of two resistors in parallel):

$$\mathbf{Y}_{XY} = (1/R) + j\omega C = 1 + j1 \text{ mS}$$

from which it follows that $\mathbf{Z}_{XY} = 1/\mathbf{Y}_{XY} = 1/(1 + j) = 0.5 - j0.5 \text{ k}\Omega$ which checks with the earlier calculation. The impedance $\mathbf{Z}_{XY}$ is in series with the

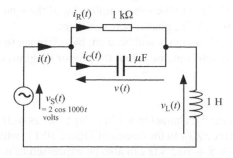

Figure 10.10 The circuit to be analysed in Example 10.3

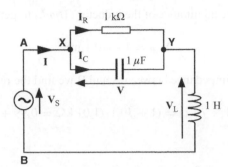

Figure 10.11   The circuit of Figure 10.10 with currents and voltages represented by complex quantities

inductor so, recalling the expression for the resistance of two resistors connected in series, we can express the total impedance between A and B as:

$$\mathbf{Z}_{AB} = \mathbf{Z}_{XY} + j\omega L = 0.5 + j0.5 \text{ k}\Omega$$

If we now connect terminals A and B to the voltage source we can calculate the resulting current:

$$\mathbf{I} = \mathbf{V}_S/\mathbf{Z}_{AB} = (2 + j0)/(0.5 + j0.5) = 2 - j2 \text{ mA}$$

Thus, the current in the inductor has an amplitude $\sqrt{(2^2 + 2^2)} = 2\sqrt{2}$ mA.

Other currents and voltages can easily be calculated. For example, the complex voltage across the inductor is $\mathbf{I}j\omega L = 2 + j2$ V, which represents a sinusoidal voltage of radian frequency 1000 and amplitude $2\sqrt{2}$ V.

Note that our selection of zero imaginary part for $\mathbf{V}_S$ is arbitrary. The choice of $\mathbf{V}_S = 0 + j2$ V would lead to the same amplitudes of the sinusoidal currents and voltages and the same relative phases between them.

The results of our AC analysis will be seen, by reference to Chapter 9, to be in agreement with the results obtained from the phasor diagram construction.

Example 10.4

Complex quantities can be characterised in polar form as well as rectangular. For example, the impedance $\mathbf{Z}_{AB}$ in the circuit of Figure 10.11, which we expressed in rectangular form as $0.5 + j0.5$ kΩ can also be expressed in polar form as having a magnitude of $0.5\sqrt{2}$ and an angle of $45°$, written as $0.5\sqrt{2}\angle 45°$. This alternative representation can be useful when working with the product or division of two complex quantities. Thus, in calculating the value of $\mathbf{I}$ we could write:

$$\mathbf{I} = \mathbf{V_S}/\mathbf{Z_{AB}} = (2\angle 0°)/(0.5\sqrt{2}\angle 45°) = 2\sqrt{2}\angle -45°$$

Since the reader may have been introduced to the concept of complex quantities only recently, a useful point to make concerns the appearance of the complex quantity j in the equations involved in AC analysis. As we have seen, to make use of the simple approach to analysis we work in the complex domain in which quantities are necessarily complex, whereas the actual currents and voltages of interest do not involve j – they are functions of time. Mistakes can easily be made if the j involved in an AC analysis is removed too soon. To take a specific instance from Example 10.1, we found that the current $\mathbf{I}$ was $0.2 - j0.2$ A. The temptation to drop the j would result in a current of zero value! Similarly, we found that $\mathbf{V_L} = 10 + j10$ V; again, 'losing' the j would result in a voltage amplitude of 20 rather than 14.14 V. The essential thing to remember is to conduct the AC analysis entirely within the complex domain until the (complex) representations of the currents and voltages of interest have been found and then, and only then, to perform the transformation back into the time domain.

Finally, it should be mentioned that the systematic method of circuit analysis introduced in the context of DC circuits in Chapter 4 is equally applicable to AC circuits. If, for example, the circuit has two unknown voltages $\mathbf{V_A}$ and $\mathbf{V_B}$, application of KCL at two nodes will result in two complex equations and hence four real equations whose solution will yield the real and imaginary parts of the two complex voltages. Such a systematic approach forms the basis of many computer programmes for the analysis of AC circuits.

10.6   Problems

Problem 10.1

For each of the 12 circuits shown in Figure P10.1, and for the stated frequency, express both the impedance and admittance between terminals A and B in complex form. Also derive the magnitude and phase of each impedance and admittance.

Problem 10.2

For the circuit of Figure P10.2, where voltages and currents are represented by complex quantities, find the values of $\mathbf{V_R}$, $\mathbf{V_C}$ and $\mathbf{I}$. Express these quantities in polar form.

If the input voltage were equivalently represented as $10 + j0$, write down by inspection (NOT by reanalysis) the new values of $\mathbf{V_R}$, $\mathbf{V_C}$ and $\mathbf{I}$.

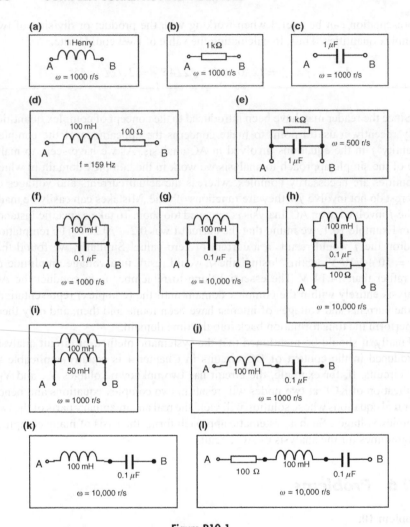

Figure P10.1

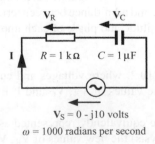

$V_S = 0 - j10$ volts

$\omega = 1000$ radians per second

Figure P10.2

Problem 10.3

For the circuit of Figure P10.3 derive an expression for the complex voltage V_O in terms of G, R, C, ω and V_{IN}. At what frequency will V_O lead V_{IN} by 135°? At this frequency what is the value of $|V_O/V_{IN}|$?

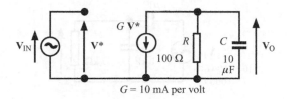

$G = 10$ mA per volt

Figure P10.3

Problem 10.4

For the circuit shown in Figure P10.4 determine the radian frequency for which the current $i(t)$ is zero whatever the amplitude of the voltage source.

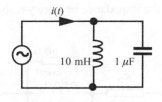

Figure P10.4

Problem 10.5

Calculate the peak-to-peak amplitude of the voltage $v_O(t)$ in the circuit of Figure P10.5. What capacitance should be connected across the inductor to ensure that $v_O(t)$ is in phase with $v_{IN}(t)$? What is then the ratio of the amplitudes of $v_O(t)$ and $v_{IN}(t)$?

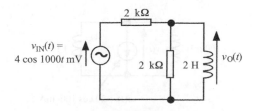

Figure P10.5

Problem 10.6

For the circuit of Figure P10.6 determine the magnitude of the complex voltage V_O. What is the phase relation between V_O and V_{IN} ?

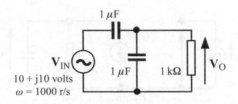

Figure P10.6

Problem 10.7

For the circuit of Figure P10.7 express the complex impedance between the terminals A and B in terms of R, R_1, L, C and the radian frequency ω. For what numerical value of ω will the impedance be purely resistive?

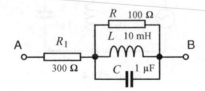

Figure P10.7

Problem 10.8

For the circuit of Figure P10.8 derive an expression for the voltage $v(t)$ appearing across the sinusoidal current source.

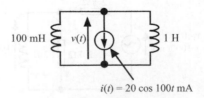

Figure P10.8

Problem 10.9

In the circuit shown in Figure P10.9, $\mathbf{V_S}$ represents a sinusoidal voltage of radian frequency ω. Derive an expression for the complex voltage $\mathbf{V}$ as a function of R, L, C and ω. Hence show that $\mathbf{V} = 0$ if $R = \sqrt{(L/C)}$. Show that if this relation between R, L and C holds, the circuit between terminals X and Y is indistinguishable electrically from a resistor of value R.

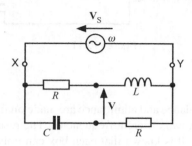

Figure P10.9

Problem 10.10

For the circuit shown in Figure P10.10 the relation between the components is $R = \sqrt{(L/C)}$. Derive an expression for the complex voltage amplification $\mathbf{V_{OUT}}/\mathbf{V_{IN}}$.

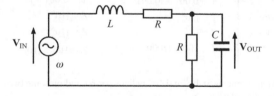

Figure P10.10

Derive an expression for the radian frequency at which the phase shift of the voltage amplification is $-90°$. Show that, at that frequency, the magnitude of the voltage amplification is $1/(2\sqrt{2})$.

Show that, at the same frequency, the input impedance 'seen' by the voltage source is $Z_{IN} = R[4/3 + j(2\sqrt{2})/3]$

Problem 10.11

A circuit discussed in Chapter 9 is reproduced as Figure P10.11. Determine the relation between the complex voltages $\mathbf{V}$ and $\mathbf{V_2}$ and check against the phasor diagram generated in Chapter 9.

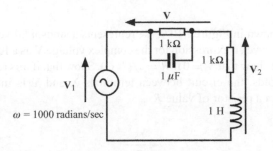

Figure P10.11

Problem 10.12

Measurements of impedance and admittance are made on five two-terminal 'black boxes', in every case at two different frequencies. The results are recorded in the table in Figure P10.12. It is known that each box can only contain either one or two components, which can be resistors, capacitors or inductors. Determine the circuit within each of the black boxes.

	Impedance Z (ohms) or Admittance Y (siemens) at $\omega = 10^3$	Impedance Z (ohms) or Admittance Y (siemens) at $\omega = 10^6$
Box 1	Z = 1000	Z = 1000
Box 2	Y = 0.001 + j0.001	Y = 0.001 + j
Box 3	Z = j10³	Z = j10⁶
Box 4	Z = 0	Z = j10⁶
Box 5	Z = 100 + j1000	Z = 100 + j10⁶

Measurements at 10^3 and 10^6 radians per second on some two-terminal black boxes

Figure P10.12

Frequency Domain Behaviour

We have seen in the last chapter how to determine the behaviour of a circuit whose source is a single-frequency sinusoid. However, many circuits are specifically designed to exhibit different behaviour at different frequencies. One example is a filter that will provide amplification for frequencies within a specified frequency range, but will provide little amplification outside that range: the example of a tuned circuit in Chapter 9 provides some indication about how this might be achieved by means of a resonant circuit. Another example is provided by the amplifier inside a portable CD player: here it is essential to provide the same amplification over a wide range of frequencies – the same for a piccolo as for a double bass – and to ensure that the inevitable drop in amplification at very low and very high frequencies is suitably placed in the frequency range (Figure 11.1) and does not limit the enjoyment of listening to a musical performance.

We also saw in the last chapter that equations describing the performance of an AC circuit contain the radian frequency ω of the sinusoidal source. For that reason it would appear that there is no need to search for a new approach to circuit analysis: one can substitute a number of values for ω and compute what circuit performance is of interest. Why, therefore, do we need a separate discussion of what is known as *frequency domain behaviour*? The reason is that for some common circuits their frequency domain behaviour exhibits interesting properties that can be of considerable help, both to the circuit designer and to anyone trying to understand the performance to be expected of a circuit.

Introductory Circuits   Robert Spence
© 2008 John Wiley & Sons, Ltd

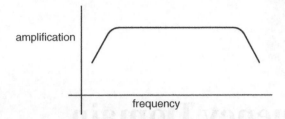

Figure 11.1   Typical form of the frequency dependence of an amplifier's gain

11.1 Asymptotic Behaviour

We choose to illustrate an approach to frequency domain analysis by using the circuit of Figure 11.2(a). It is a simple circuit, but recall that the combination of voltage source and resistor could be the Thévenin equivalent of a large resistive circuit. The currents and voltages of interest are sinusoidal functions of time. Following the analysis procedure established in Chapter 10, we transform the original circuit to obtain another identical in form (Figure 11.2.b), but with all currents and voltages represented by complex quantities.

For the circuit of Figure 11.2(a) a typical question is 'how does the ratio V_C/V_S of "input" and "output" voltage amplitudes vary with frequency?' To find the answer we carry out the following analysis for the circuit of Figure 11.2(b). From Ohms law:

$$\mathbf{V_R} = R\mathbf{I} \qquad (11.1)$$

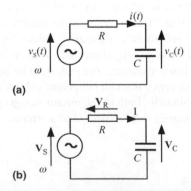

(a)

(b)

Figure 11.2   (a) The circuit whose currents and voltages are of interest; (b) the same circuit with currents and voltages represented by complex values

From 'Ohm's law' for a capacitor:

$$\mathbf{V}_C = (1/j\omega C)\mathbf{I} \tag{11.2}$$

From KVL:

$$\mathbf{V}_S = \mathbf{V}_R + \mathbf{V}_C = \mathbf{I}[R + 1/(j\omega C)] \tag{11.3}$$

Combining the last two equations we find:

$$\mathbf{V}_C/\mathbf{V}_S = (1/j\omega C)/[R + 1/j\omega C] \tag{11.4}$$

which, of course, we could have derived directly using the voltage divider principle. Simple rearrangement of the last equation gives

$$\frac{\mathbf{V}_C}{\mathbf{V}_S} = \frac{1}{1 + j\omega CR} \tag{11.5}$$

Many circuit properties vary with frequency in this way.

That, however, is not the end of the story, because much is to be gained by the way in which the function in Equation (11.5) is plotted.

If, as often happens, we are interested in how the magnitude of the voltage amplification $|V_C/V_S|$ varies with frequency we could simply plot $|\mathbf{V}_C/\mathbf{V}_S|$ versus ω on a conventional graph (Figure 11.3). There is nothing wrong with that graph: we can see, for example, that the amplification tends to unity at low frequencies and to zero at high frequencies. There is, however, an equivalent, but much more informative way of illustrating and gaining insight from Equation (11.5).

Instead of plotting $|\mathbf{V}_C/\mathbf{V}_S|$ versus ω we shall now plot $\log_{10}|\mathbf{V}_C/\mathbf{V}_S|$ versus $\log_{10}\omega$. This would appear to be more complicated – but it isn't, as we shall see – and seems to require us to compute new quantities to plot – but it doesn't.

We start by looking at high values of the radian frequency ω. To be specific, we examine Equation (11.5) for values of ω which are much greater than $1/CR$. For

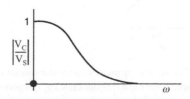

Figure 11.3 A plot of $|V_C/V_S|$ for the circuit of Figure 12.2 (linear scales)

those values of ω we can state that

$$|\mathbf{V_C}/\mathbf{V_S}| \text{ tends to } 1/\omega CR \text{ as } \omega \text{ tends to infinity} \qquad (11.6)$$

If we plot

$$|\mathbf{V_C}/\mathbf{V_S}| = 1/\omega CR \qquad (11.7)$$

using the axes $\log_{10}|\mathbf{V_C}/\mathbf{V_S}|$ and $\log_{10}\omega$ we get the straight line shown in Figure 11.4(a). For convenience the corresponding values of $|\mathbf{V_C}/\mathbf{V_S}|$ and ω are shown alongside the log scales. A check shows that the amplification $|\mathbf{V_C}/\mathbf{V_S}|$ decreases by ten times for a ten times increase in ω, which agrees with Equation 11.7.

Three points should be noted. First, one can purchase what is called log–log graph paper, so designed that there is no need to take logs: in Figure 11.4(a) it is the scale to the left of the vertical axis we shall use, because a point on that scale (which is more meaningful to us) corresponds to the correct point on the log scale. The same comment applies to the horizontal axis. Second, the plot of Figure 11.4(a)

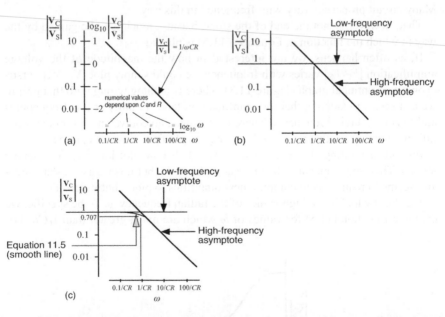

Figure 11.4   Plot of $|V_C/V_S|$ versus frequency ω for the circuit of figure 11.2 using logarithmic scale. (a) The asymptote for high frequencies; (b) the asymptote for low frequencies is added: logarithmic scales are still employed but values not shown; (c) the smooth curve of Equation (11.5) is added, showing a value of $|V_C/V_S|$ of $1\sqrt{2}$ at the frequency $\omega = 1/CR$

provides an *approximation* to the relation (11.5). The approximation is excellent for very high values of ω, but decreases in accuracy as ω approaches the value $1/CR$. Third, although the axes in Figure 11.4(a) are continuous they have purposely been drawn to avoid any crossing which might suggest a zero value for the $|\mathbf{V_C}/\mathbf{V_S}|$ and ω axes. Because we are using log scales the zero values of $|\mathbf{V_C}/\mathbf{V_S}|$ and ω appear at minus infinity!

We now turn our attention to values of ω much less than $1/CR$. We can state from examination of Equation (11.5) that

$|\mathbf{V_C}/\mathbf{V_S}|$ tends to a value of unity as ω tends to zero

If we now add a plot of

$$|\mathbf{V_C}/\mathbf{V_S}| = 1$$

to Figure 11.4(a) – again plotting $\log_{10}|\mathbf{V_C}/\mathbf{V_S}|$ versus $\log_{10}\omega$ – we obtain the plot shown in Figure 11.4(b). In this plot we have intentionally not shown numerical values on the log axes to emphasize the fact that we do not need to take logs – the printer of the log–log paper has done that for us! Like the straight line drawn for high values of ω, the line we have just added is also an approximation to Equation (11.5): in each case the actual value of $|\mathbf{V_C}/\mathbf{V_S}|$ asymptotes to these lines at low and high values of ω. It is for this reason that the two straight lines are referred to as the low-frequency and high-frequency *asymptotes* of equation 11.5. We note that they intersect at $\omega = 1/CR$.

The actual relation between $|\mathbf{V_C}/\mathbf{V_S}|$ and ω is, of course, a smooth and continuous one as seen from Equation (11.5). If we now add a plot of Equation (11.5) to what we have already we obtain Figure 11.4(c). It is easy to see from Equation (11.5) that at the frequency $\omega = 1/CR$ the value of $|\mathbf{V_C}/\mathbf{V_S}|$ is $1/\sqrt{2}$.

It may seem surprising that the continuous plot of Equation (11.5) is not always of primary interest. The reason is that, when designing a circuit, the primary interest may well be the *general positioning* of the asymptotes. For example, if the circuit of Figure 11.2(a) is that of a filter, designed to allow low frequencies in the source voltage to pass essentially undiminished in amplitude to the output, but to attenuate high frequencies, then the plot of Figure 11.4(b) shows that the amplification experienced by frequencies up to a value of about $1/CR$ will be approximately unity, whereas amplification decreases significantly beyond that frequency. Moreover, it is clear to the circuit designer that it is the product CR which determines when the amplification decreases, and also that variation in C and/or R will not affect the low frequency asymptote. For larger circuits containing many components whose value has to be chosen by a designer, such information is extremely valuable, and it will often be the case that, in the early stages of design, the designer will focus on the position – and positioning – of the asymptotes rather than a precise plot of a circuit property.

Example 11.1

The circuit of Figure 11.5(a) contains a resistor and an inductor connected to a sinusoidal voltage source. We are asked to sketch a plot of the magnitude of the ratio $|V_2/V_1|$ of the amplitudes V_1 and V_2 of the input and output voltages $v_1(t)$ and $v_2(t)$ versus the radian frequency ω of the source.

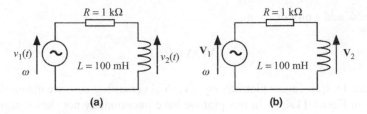

(a)   **(b)**

Figure 11.5   (a) The circuit of Example 11.1; (b) the same circuit with sinusoidal voltages represented by complex voltages

Following the analysis procedure discussed in Chapter 10 we redraw the circuit, but show the complex representations of voltage, as in Figure 11.5(b). Using the voltage divider principle we can write:

$$\mathbf{V}_2/\mathbf{V}_1 = j\omega L/(R + j\omega L) = \cfrac{1}{1 - j/\left(\cfrac{\omega L}{R}\right)} \tag{11.8}$$

which has a similar form to Equation (11.5).

The low-frequency asymptote relevant to frequencies less than $\omega = R/L$ is given by

$$|\mathbf{V}_2/\mathbf{V}_1| = \omega L/R$$

which describes, when log scales are employed for $|\mathbf{V}_2/\mathbf{V}_1|$ and ω, a straight line whose slope is unity if the same log scales are used for both variables (see Figure 11.6). As the frequency ω increases beyond R/L the value of $|\mathbf{V}_2/\mathbf{V}_1|$ approaches the high-frequency asymptote given by

$$|\mathbf{V}_2/\mathbf{V}_1| = 1$$

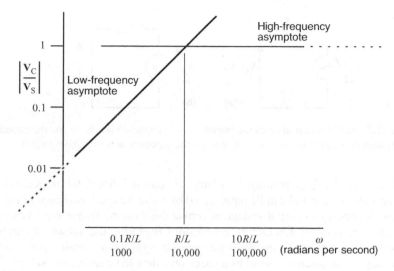

Figure 11.6   The asymptotes associated with the voltage ratio $|V_1/V_2|$ in the circuit of Figure 11.5

which is also plotted in Figure 11.6. The two asymptotes intersect at

$$\omega L/R = 1$$
$$\text{or} \quad \omega = R/L$$

Since $|\mathbf{V}_2/\mathbf{V}_1|$ is identical with $|V_2/V_1|$, the plot of Figure 11.6 provides the requested information about the performance of the circuit of Figure 11.5(a).

The continuous plot of Equation (11.8) has intentionally not been shown in order to emphasize the considerable value of the asymptotes on their own. For the given values of R and L (such that $R/L = 10^4$r/s) the ω axis has been assigned corresponding values.

11.2 Extreme Frequencies

For a circuit containing sinusoidal sources a useful part-check on its performance can be carried out by examining that performance at the extreme frequencies of zero and infinity. Fortunately, this check can easily be carried out, often by inspection. Take, for example, the circuit of Figure 11.2(a). At zero frequency the impedance of the capacitor is infinite, allowing the circuit to be represented as shown in Figure 11.7(a), from which it is clear that the magnitude of the voltage

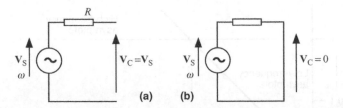

Figure 11.7 (a) At zero frequency the impedance of a capacitor is infinite and the capacitor acts as an open-circuit; (b) at infinite frequency the capacitor acts as a short-circuit

amplification, $|V_C/V_S|$, is unity. Similarly, when ω is infinite, the impedance of the capacitor is zero and can be represented by a short-circuit as shown in Figure 11.7(b), leading to an output voltage of zero at this extreme frequency. Even with a much more complicated circuit, conclusions regarding the values of currents and voltages at extreme frequencies are fairly straightforward, avoiding a detailed analysis, and can often be useful as a check on a detailed circuit analysis.

11.3 Opamp Limitations

In Chapter 7 we employed an opamp within an inverter circuit (repeated as Figure 11.8) and discovered that the magnitude of the voltage amplification was equal to the ratio of the two resistors R_2/R_1. For illustration we chose $R_2 = 10R_1$, but acknowledged that the resulting amplification of ten times seemed disappointingly low compared with the voltage gain (typically 10^4 to 10^6) provided by the linear region of the opamp. One reason for making that choice of R_2/R_1 can now be discussed, because we now have the analytical basis for explanation.

First, it is a fact that the very high voltage gain provided by the linear region of the opamp remains high only if the frequency of the voltages V_1 and V_O remains

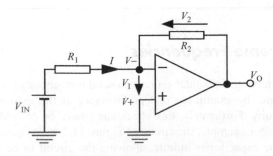

Figure 11.8 The inverter circuit

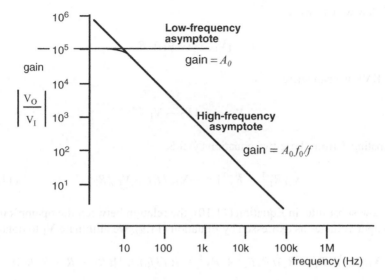

Figure 11.9 Typical low- and high-frequency asymptotes of the gain of an opamp

relatively low. How low can be quite surprising (Figure 11.9): it would not be unusual to find the gain beginning to reduce at frequencies as low as 10 Hz!

To investigate what effect this rather severe limitation has on a circuit containing such an opamp we need to characterize the sort of performance shown in Figure 11.9 analytically. Fortunately, it turns out that the performance illustrated in Figure 11.9 can be expressed as

$$\frac{\mathbf{V}_O}{\mathbf{V}_I} = A_0 \left[\frac{1}{1 + jf/f_0} \right] \qquad (11.9)$$

where A_0 is the low-frequency gain of the opamp. For the opamp whose performance is illustrated in Figure 11.9 the value of f_0 is 10 Hz and the value of A_0 is 10^5. Thus, the functional dependence of $\mathbf{V}_O/\mathbf{V}_I$ upon frequency is the same as that of the RC circuit of Figure 11.2(a) and expressed in Equation (11.5). Figure 11.9 additionally shows and characterizes the low- and high-frequency asymptotes of Equation (11.9).

With the help of Equation (11.9) we can now undertake a new analysis of the inverter circuit, repeated in Figure 11.8. However, to take into account frequencies for which the opamp's gain is not high, we must discard the notion of a virtual short circuit ($\mathbf{V}_I = 0$) and use instead the Equation (11.9) containing $\mathbf{V}_I$. Thus, by

Ohm's law we can write

$$I = [\mathbf{V}_{IN} - (-\mathbf{V}_1)]/R$$

From KVL we can write

$$\mathbf{V}_O = -IR_2 - \mathbf{V}_1$$

Eliminating I from these two equations yields

$$\mathbf{V}_1[R_1^{-1} + R_2^{-1}] = -\mathbf{V}_{IN}/R_1 - \mathbf{V}_O/R_2 \tag{11.10}$$

If we now substitute, in Equation (11.10), the relation between the opamp's input and output voltages as expressed by Equation (11.9), we eliminate $\mathbf{V}_1$ to obtain:

$$\mathbf{V}_O[A_0^{-1}\{(R_1 + R_2)/R_1R_2\} + R_2^{-1} + j(f/f_0)A_0^{-1}\{(R_1 + R_2)/R_1R_2\}]$$

If we make the reasonable assumptions that $A_0 \gg 1$ (A_0 is typically at least 10^3) and $R_2/R_1 \gg 1$ (because that is a typical choice to achieve reasonable voltage amplification) we can write

$$\frac{\mathbf{V}_O}{\mathbf{V}_{IN}} = -\frac{R_2}{R_1}\left[\frac{1}{1 + jf/f_0'}\right] \tag{11.11}$$

where $f_0' = f_0[A_0/(R_2/R_1)]$.

A first glance at Equation (11.11) shows that, at low frequencies, the voltage amplification is equal to $-R_2/R_1$, as we would expect. Moreover, as the frequency f increases, the amplification decreases as again we might expect. But what is interesting is the expression for f_0', the 'cut-off' frequency of the inverter. We see that the product of (R_2/R_1), (which is the magnitude of the low-frequency amplification) and f_0' (the new 'cut-off' frequency) is a constant A_0f_0, *defined solely by the opamp*: A_0 is its low-frequency amplification, and f_0 is its (low) cut-off frequency. We have, in fact, discovered a trade-off (something common to all engineering design) in this case showing that the greater the low-frequency amplification (R_2/R_1) of the inverter, the smaller is its cut-off frequency f_0'. It is interesting to show this trade-off on a plot (Figure 11.10) of the opamp's frequency characteristic, using some examples of designs involving different values of R_2/R_1, the magnitude of the low-frequency gain.

This analysis has provided one answer to the question posed in Chapter 7 regarding the inverter: 'why employ such low vales of R_2/R_1 (and hence amplification) when the opamp has such a high gain in its linear region?'

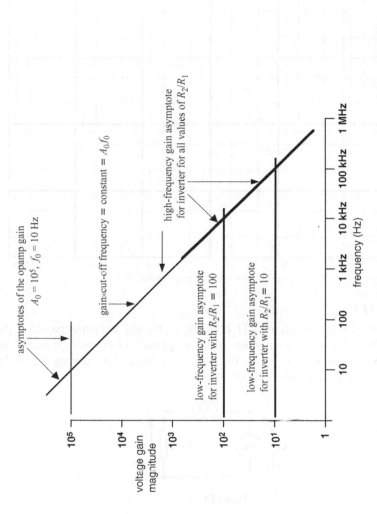

Figure 11.10 The frequency dependence of voltage gain for the inverter circuit and for the opamp within it, showing a trade-off between gain magnitude and bandwidth for the inverter. The product of the inverter's gain and cut-off frequency is defined by the opamp

11.4   Problems

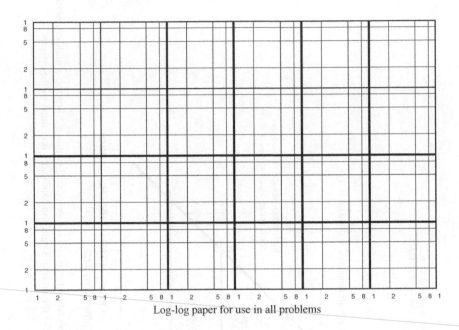

Log-log paper for use in all problems

Problem 11.1

For the circuit shown in Figure P11.1 sketch, on the log–log paper provided, the low- and high-frequency asymptotes of the magnitude of the voltage amplification $|V_2/V_1|$ where V_1 and V_2 are the complex voltages representing sinusoidal voltages. Put numerical values on the axes.

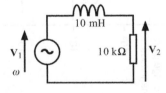

Figure P11.1

Problem 11.2

Sketch the low- and high-frequency asymptotes of the magnitude of the current ratio I_2/I_1 for the circuit shown in Figure P11.2. I_1 and I_2 are complex quantities representing sinusoidal currents. Use the log–log paper provided and label the axes appropriately.

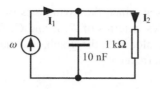

Figure P11.2

Problem 11.3

In the active filter shown in Figure P11.3 V_{in} and V_{out} are complex quantities representing sinusoidal voltages. Assuming that the operation of the opamp remains in its linear high gain region, choose values for R, R_A and C to ensure that the voltage gain V_{out}/V_{in} has a magnitude of 15 at low frequencies and a 'cut-off frequency' of 100 kHz. On the log-log paper provided sketch the gain asymptotes of the filter you have designed.

Which components affect the gain at low frequencies? Which components determine the cut-off frequency?

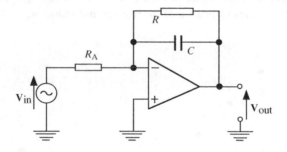

Figure P11.3

Problem 11.4

The series connection of a resistor, a capacitor and an inductor is shown in Figure P11.4.

First, derive an expression for the complex impedance Z between terminals A and B at a radian frequency ω in terms of R, L, C and Ω.

Next, using the numerical component values given, plot on the log–log paper provided the asymptote (NOT the actual values) of $|Z|$ for high frequencies, where the inductor is dominant. Then plot the asymptote of $|Z|$ for low frequencies, where the capacitor is dominant. Finally, under the assumption that there are frequencies for which the resistance is dominate, calculate and plot the asymptote to which $|Z|$ tends at those frequencies.

For the two frequencies at which the asymptotes intersect identify the corresponding radian frequencies in terms of R, L and C. For these two frequencies calculate the value of $|\mathbf{Z}|$, plot it on the graph, and sketch (don't calculate) the actual variation of $|\mathbf{Z}|$ with frequency.

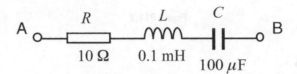

Figure P11.4

Problem 11.5

Measurements have been made on a circuit comprising the series connection of a resistor, a capacitor and an inductor. The measured magnitude of the impedance of the series connection is plotted against frequency in Figure P11.5. Estimate, with explanation, the capacitance of the capacitor, the inductance of the inductor and the combined series resistance of the resistor and inductor.

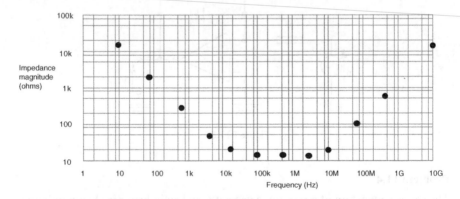

Figure P11.5

Problem 11.6

Figure P11.6(a) shows the circuit of a filter using an operational amplifer. Figure P11.6(b) shows the specifications placed by a customer on a filter. Any design falling within the shaded area is acceptable.

Derive expressions for the low-frequency, mid-frequency and high-frequency asymptotes of the magnitude of the voltage gain $\mathbf{V_O}/\mathbf{V_{IN}}$, where $\mathbf{V_O}$ and $\mathbf{V_{IN}}$ are complex voltages representing the sinusoidal input and output voltages of the circuit.

If possible choose values of the two resistors and two capacitors such that the circuit will satisfy the customer's specification.

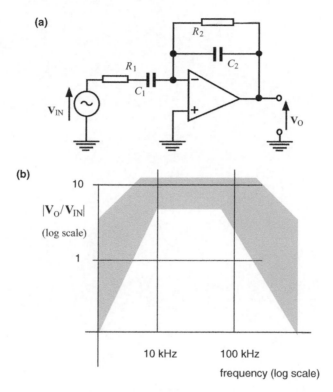

Figure P11.6

Overview: The Analysis of Change

The last type of circuit behaviour we shall examine is concerned with change.

One example of change that is of considerable interest is when you connect a load to a power supply, because the resulting change in power supply voltage may be significant and unacceptable. Sometimes these changes can be quite large (and, for example, affect other circuits connected to a power supply) so we must be able to predict them.

Another extremely important example of change is that which is inherent when a sinusoidal voltage source is applied to a circuit. We have of course dealt with this situation in the AC chapters, but only for linear circuits. Many circuits of practical importance (e.g., amplifiers) are nonlinear because they contain components such as diodes and transistors, but are operated at sufficiently small amplitudes that essentially linear operation results. Nevertheless we need to know how to handle such nonlinear components under these circumstances, by using what is known as small-signal analysis.

Fortunately we shall find that the equations relating changes in currents and voltages obey the same sorts of laws as do simple DC currents and voltages, so that analysis is reasonably straightforward.

In the course of the following chapters we shall meet two new components – the Zener and exponential diodes – and see how they can be usefully employed in circuits.

Introductory Circuits   Robert Spence
© 2008 John Wiley & Sons, Ltd

Change Behaviour

Many of the circuits we have examined have either contained ideal voltage sources or have indicated that a voltage supply is required to ensure that a circuit performs as required. The problem remains as to how one obtains something approximating to an ideal voltage source.

A number of solutions are possible. A familiar one is to use batteries. Another is to exploit the fact that the voltage across a forward-biased diode is relatively insensitive to the current through it (Figure 12.1a) so that, within a circuit (Figure 12.1b), if sufficient current can be made to flow through a series connection of diodes, the voltage across that connection is relatively stable. Usually, the stabilized voltage will be an integer multiple of 0.7 V in view of the nature of the diode characteristic.

12.1 Voltage Stabilization

To introduce an important concept we consider a third approach to the generation of an essentially constant voltage. It is made possible by the unique characteristic of the Zener diode (Figure 12.2). Over a substantial part of its current range (called the Zener breakdown region) its voltage is essentially independent of the current through it. Therefore, if we can ensure that its current stays within that current range, the voltage across it will approximate to that of an ideal voltage source.

A very simple circuit that exploits this property of a Zener diode is shown in Figure 12.3. At the left we have a 15 V ideal source which we shall for the moment assume to be of constant value. On the right we have a resistor R (the 'load') representing a piece of equipment that requires an essentially constant voltage of 5 V. We shall also assume that this load resistance is liable to vary over quite a

Introductory Circuits   Robert Spence
© 2008 John Wiley & Sons, Ltd

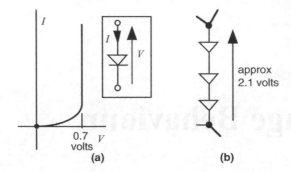

Figure 12.1 A diode characteristic, and the use of diodes to establish a fixed voltage within a circuit

large range. In the middle we have a resistor R_S and a Zener diode whose Zener voltage V_Z (Figure 12.2) is 5 V.

To see how the circuit works consider the situation in which the current I_L through the load is 100 mA (in other words, $R = 50\,\Omega$), and that the voltage across the load is 5 V. Assume we need at least 1 mA through the Zener to keep the diode in its breakdown region. By KCL the current through the resistor R_S will be 101 mA so that the value of R_S is, by Ohm's law, $(15 - 5)/101$ mA $= 99\,\Omega$. That choice of R_S will ensure that 5 V will appear across R. The operating point of the Zener is shown as A in Figure 12.2.

Now suppose that the load resistance is increased so that $I_L = 50$ mA. If the voltage across the load is still 5 V, such that the voltage across R_S is still 10 V, the same current as before (101 mA) will flow into the combination of load resistance and diode. Since the load current is 50 mA this means that we now have 51 mA flowing in the diode (operating point B in Figure 12.2). Finally, we suppose that the load resistance is disconnected, so that $I_L = 0$. Then, the whole of the 101 mA flows through the diode, such that its operating point C is as shown in Figure 12.2. Overall we have a constant current (of 101 mA) diverted to the load and Zener diode according to the value of the load resistance R.

We can draw two conclusions from this exercise. First, that a Zener diode is capable of maintaining an essentially constant voltage across a resistive load, even when the current drawn by that load varies considerably. This ability fails when an attempt is made to draw too much current through the load (>100 mA in the example discussed). Second, we note that because the Zener characteristic does not quite correspond to a constant voltage in the breakdown region, the voltage V across the load, which is nominally 5 V, actually decreases slightly as more current is taken by the load (from point C to point A in Figure 12.2). The extent to which the voltage decreases depends upon the slope of the Zener characteristic. We shall see how to calculate this decrease in the next section.

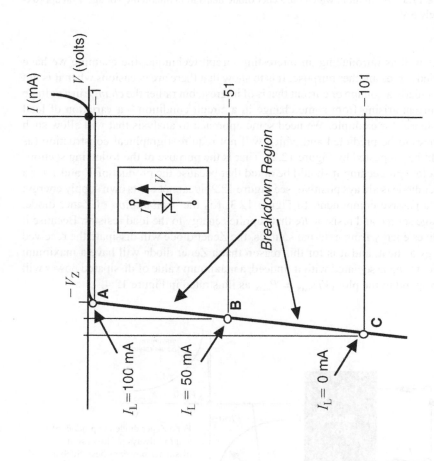

Figure 12.2 The voltage~current characteristic of a Zener diode. Operation in the breakdown region is nondestructive

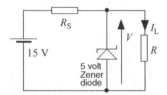

Figure 12.3   A circuit in which the Zener diode helps to maintain the voltage V at approximately 5 V

As well as introducing an interesting circuit technique, the example we have studied serves another purpose. It is to show that there are occasions when it is not particularly a voltage or current that is of interest, but rather the *change* in a voltage or current arising from some change in a circuit condition – a variation of load resistance, for example. We need some approach to analysis that will allow such changes to be predicted and which will not require a graphical construction (as might be suggested by Figure 12.2). That is the purpose of the following section.

Before proceeding it should be noted that because the product of V and I for a Zener diode is always positive (see Figure 12.2) it cannot on its own supply energy: it is a passive component. In Figure 12.3 it is the *combination* of Zener diode, voltage source and resistor R_S that supplies energy to the load resistor. Because it receives energy from external sources, the Zener diode will dissipate the received energy as heat, and it is for this reason that a Zener diode will have a maximum power rating associated with it. Indeed, a maximum value of dissipated power will correspond to the plot $(VI)_{max} = P_{max}$ as illustrated in Figure 12.4.

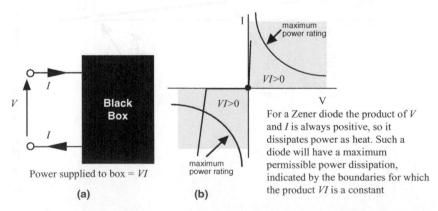

Power supplied to box = VI

(a)

(b)

For a Zener diode the product of V and I is always positive, so it dissipates power as heat. Such a diode will have a maximum permissible power dissipation, indicated by the boundaries for which the product VI is a constant

Figure 12.4   (a) The condition for a component to be passive; (b) a Zener diode is passive: it cannot supply energy

12.2   The Analysis of Change

Connections

In our search for a method of analysis we follow the same route as for DC and AC circuits. We first explore to see if Kirchhoff's current law applies to the connections between components. Figure 12.5(a) shows part of a circuit in which currents I_1, I_2 and I_3 flow into a particular node. We know from KCL that $I_1 + I_2 + I_3 = 0$. Now imagine that changes take place in the external circuit and cause changes in these currents (Figure 12.5b) so that they become $I_1 + \Delta I_1$, $I_2 + \Delta I_2$ and $I_3 + \Delta I_3$. Again, these currents must sum to zero, and therefore we can say that

$$\Delta I_1 + \Delta I_2 + \Delta I_3 = 0 \tag{12.1}$$

which provides an illustration of the fact that *KCL applies to changes in currents* as well as to the currents themselves. We note that there is no restriction on the magnitude of the changes.

A similar example can illustrate the fact that *Kirchhoff's voltage law also applies to changes in voltage.*

Component relations

Having looked at the effect of *connections* on voltage and current changes we now examine the relations imposed between changes in voltage and current by *components*: obviously it would be fortunate if we can find Ohm's law-type relations!

We begin with the linear resistor, described by Ohm's law $V = RI$. It follows that changes in voltage ΔV and changes in current ΔI must also be related by a

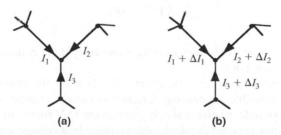

Figure 12.5   (a) The currents flowing into a node and; (b) new values of those currents following a change somewhere in the external circuit

similar law:

$$\Delta V = R\Delta I \qquad (12.2)$$

Next we examine an ideal voltage source of constant value. Clearly, if a voltage is constant, its change must be zero. Thus, for an ideal voltage source,

$$\Delta V = 0 \qquad (12.3)$$

Similarly, an ideal current source of constant value is characterised by zero changes in current:

$$\Delta I = 0 \qquad (12.4)$$

We now turn to a more complex component, the Zener diode. It may seem odd that we might be hoping for a linear model of such a nonlinear component, but if we concentrate our attention on that part of the Zener diode's characteristic that we are making use of, we easily find a linear relation. Thus, whereas the Zener's actual characteristic is shown in Figure 12.6(a), we model it by the linear relation shown in Figure 12.6(b): we know that (provided we design our circuit correctly!) the currents expected will always lie in the Zener breakdown region. The linear model in Figure 12.6(b) is described by

$$V = IR_Z - V_Z$$

so that changes in V and I are related by

$$\Delta V = \Delta IR_Z \qquad (12.5)$$

In other words, as far as *changes* in voltage and current are concerned, the Zener 'looks like' a resistor R_Z.

Two more components must be examined. They are the changes in current and voltage responsible for causing current and voltage changes elsewhere in a circuit. If, for example, a voltage supply (such as the 15 V source in Figure 12.3) is unreliable and may vary, we can model this variation by a voltage source. Similarly, if the current drawn by a resistive load changes, we can model that effect by a current source equal to the change.

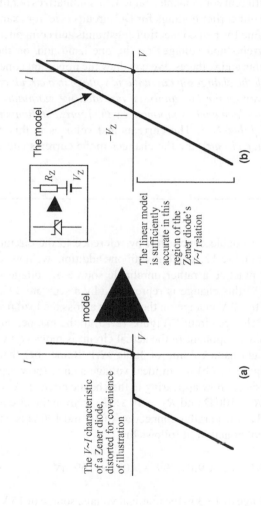

Figure 12.6 (a) The voltage~current characteristic of a Zener diode; (b), a linear model adequately representing the characteristic *in the region in which it is intended to operate*

Change analysis

A remarkable result follows from our discovery that Kirchhoff's laws apply to changes in current and voltage, and from our derivation of the relations between ΔI and ΔV imposed by different components. Table 12.1 summarizes our findings, but places them alongside our earlier findings for DC circuits (see, for example, Table 3.1). We see that the same form of connection constraints and component relations apply both to DC currents and voltages on the one hand and, on the other, to *changes* in those currents and voltages. We can draw the important conclusion that *an analysis to establish the changes in current and voltage in a circuit can proceed in exactly the same way as for DC analysis, provided we maintain component interconnections and replace each component by the representation shown in the right-most column of Table 12.1.* The currents and voltages in this new circuit (we'll call it the '*change circuit*') are the changes in the currents and voltages of the actual circuit.

Example 12.1

We can illustrate this valuable conclusion by reference to the circuit discussed above and repeated as Figure 12.7(a), but with one addition: we now assume that the 15 V source is in practice a rather unreliable source of voltage which may drift by as much as 2 V: this change is represented by a separate 2 V source. To calculate the effect of this 2 V change on the voltage across the load resistance we simply draw the new 'change circuit' (Figure 12.7b) in the manner suggested by Table 12.1, so that each component in the actual circuit is replaced by its change equivalent from the right-hand column of Table 12.1. Remember that the Zener diode is modelled (Figure 12.6b) by an ideal voltage source (now replaced by a short-circuit) and a resistor (now appearing in the change circuit). As an example we shall assume that $R = 100 \, \Omega$ and $R_Z = 10 \, \Omega$. To analyse the circuit of Figure 12.7(b) we first note that the parallel connection of 10 and 100 Ω is equivalent to 9.09 Ω. By voltage divider action, it follows that

$$\Delta V = 2 \times 9.09/(99 + 9.09) = 169 \, \text{mV}$$

Thus, a percentage change of 13.3 in the nominal voltage source of 15 V is reduced to a percentage change of 3.38 in the voltage V.

It is important to note that the change circuit of Figure 12.7(b) relates only *changes* in current and voltage. The 15 V appearing in the actual circuit does not appear anywhere in the change circuit; neither does the 5 V we wish to achieve across the Zener nor the 101 mA flowing through R_S.

Table 12.1 The component and connection relations for changes in current and voltage have the same form as do the same relations for DC currents and voltages

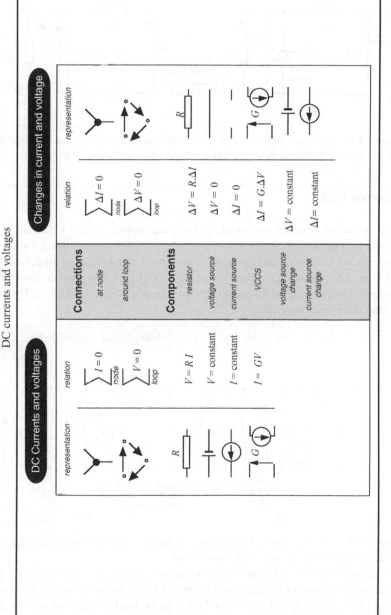

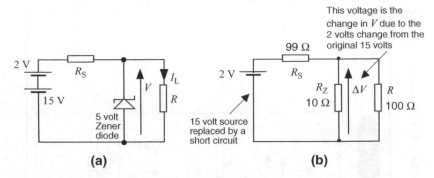

Figure 12.7   An actual circuit (a) and, (b), the change circuit relevant to the calculation of the effect of a 2 V change in the voltage source whose nominal value is 15 V

Example 12.2

As a second illustration we calculate the effect on the voltage V in the circuit of Figure 12.3 of an increase in load current of 10 mA, now assuming that the 15 V source is not liable to change. The relevant circuit and its change model are shown in Figure 12.8. In the change model a current of 10 mA is applied to three resistors in parallel whose equivalent resistance can be calculated to be 8.33 Ω. The change in load voltage is therefore

$$\Delta V = -10 \times 10^{-3} \times 8.33 = -0.0833 \text{ V or } -83.3 \text{ mV}$$

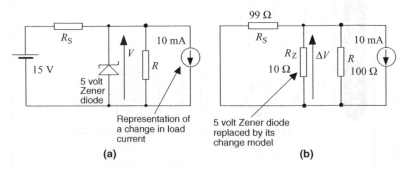

Figure 12.8   An actual circuit (a) and, (b), the change circuit relevant to the calculation of the effect of a 10 mA change in the load current

This represents a 1.6% decrease in the nominal value of 5 V for V. That a decrease was expected is suggested in Figure 12.2 by the movement of the operating point of the Zener from C to B and then A as the load current was increased.

Changes in nonlinear circuits

One limitation of the analysis approach discussed in this chapter is the assumption of linear components. In the next chapter we see how the method can easily be extended, under certain restrictions, to circuits containing nonlinear components.

12.3 Problems

Problem 12.1

The circuit of Figure P12.1 is designed to provide, at the output terminal B, a voltage approximating closely to $10\,V$. The Zener diode has a Zener voltage of $10\,V$ and an internal resistance of $10\,\Omega$.

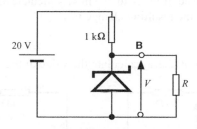

Figure P12.1

What is the approximate minimum value for R for which the circuit is able to maintain the output voltage at $10\,V$?

Calculate, to a good approximation, the current in the diode, and the value of the voltage V, for three values of the load resistor R: 5, 2 and 0.5 kΩ.

What change occurs in the voltage V if the supply voltage (nominally 20V) changes from 20 to 22 V when $R = 20\,k\Omega$?

Problem 12.2

The circuit of Figure P12.2 must be designed to provide a stabilized voltage V of approximately 6 V for load currents I less than or equal to 10 mA. The Zener diode shown has a Zener voltage of 6 V, an internal resistance of $10\,\Omega$ and should conduct a current of at least 1 mA to ensure operation in the Zener region.

Choose and justify a value of the resistance R that will ensure that the circuit meets the requirements set out above.

If, with your chosen value of R, the load resistance R_L is made equal to $3\,k\Omega$, what is the approximate value of the voltage V?

If, with your chosen value of R, the load resistance is made equal to $400\,\Omega$, what is the approximate value of the voltage V?

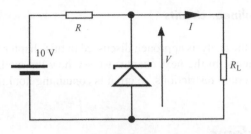

Figure P12.2

If, when $I = 5\,\text{mA}$, the value of the voltage source is increased from 10 to 12 V, what is the resulting increase in the voltage V?

Assume that the voltage source is restored to a fixed value of 10 V. The load current is now increased from 5 to 5.5 mA. Calculate the approximate value (including the sign) of the resulting change in V.

Problem 12.3

For the circuit of figure P12.3(a) calculate the value of the voltage V.

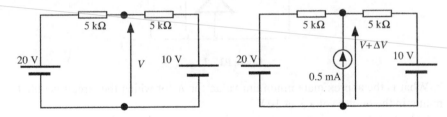

Figure P12.3

The circuit now undergoes the change indicated in Figure P12.3(b): a current source of 0.5 mA is introduced. Calculate the change ΔV in the voltage V. Check your answer by carrying out a conventional analysis of the new circuit of figure P12.3(b).

The 20 and 10 V sources in Figure P12.3(a) are now replaced, respectively, with 50 and 2 V sources, leading to a new value of the voltage V. Without carrying out any calculation, but purely by reasoning based on the earlier calculation, state the change ΔV in V that will occur when a 0.5 mA current source is connected in the same manner as in Figure P12.3(b).

13

Small-signal Analysis

In the last chapter we saw how we could predict the effect of changes in a linear circuit. We now extend our discussion to accommodate two new features. First, many useful circuits contain *nonlinear* components such as diodes, so that the approach to change analysis we have developed, based on the property of *linearity*, is no longer applicable without modification. Second, whereas the examples we have discussed referred to 'unwanted changes', the changes in current and voltage may often be extremely useful in that they represent information. For example, the varying voltage from a CD player may represent an orchestral concert, and our circuit may be an amplifier that accepts these changes in voltage and generates larger voltage changes – but of the same waveform – to apply to a loudspeaker. There is therefore a need to be able to predict the performance of a circuit that contains one or more nonlinear components and whose currents and voltages are varying in response to a source of varying voltage.

In this chapter we also take the opportunity to consider circuits in which the voltage at any node may be composed of two parts: a constant voltage to which is added a time-varying voltage, the latter usually being a useful signal of some sort. To distinguish the various components of a voltage at node Y we typically denote the actual voltage as $v_Y(t)$, its (constant) average value as V_Y and its time-varying component of average value zero by $v_y(t)$, so that

$$v_Y(t) = V_Y + v_y(t) \qquad (13.1)$$

13.1 The Extension of Change Analysis

If we are to handle the presence of nonlinear components, only a simple modification is required to the method of change analysis developed in Chapter 12. To take

Introductory Circuits Robert Spence
© 2008 John Wiley & Sons, Ltd

an example, Figure 13.1(a) shows the voltage–current characteristic of a diode. It is certainly nonlinear, but we could take the view that, if changes in current and voltage are *sufficiently small,* then there will be an approximately linear relation between them. However, it is clear from Figure 13.1(a) that the value of $\Delta I/\Delta V$ depends upon the values of current and voltage around which the changes occur, values we call the *quiescent, bias* or *average* values. For example, if the diode is operating at point **A** the ratio $\Delta I/\Delta V$ is smaller than at point **B**. By contrast, for a linear resistor (Figure 13.1b), the ratio $\Delta I/\Delta V$ is independent of the current and voltage around which the changes occur. Thus, the method of change analysis developed in Chapter 12 is perfectly valid for circuits containing nonlinear components provided that we: (1) restrict the magnitude of the changes so that we can use a linear approximation to part of a nonlinear function; and (2) employ as a change model for each nonlinear component a resistance corresponding to the value of $\Delta I/\Delta V$ at the operating point. Thus, the right-hand column of Table 12.1 undergoes the minor, but important addition highlighted in Figure 13.2. When changes are small the value of $\Delta I/\Delta V$ is called the *incremental conductance* and denoted by g_d: its reciprocal r_d is called the *incremental resistance.*

It is because we have to restrict the magnitude of the changes that we refer to *small-signal operation.* The term 'signal' is used because the small variations in current and voltage usually represent a signal of some sort, such as speech or music. The obvious question 'How small is *small*?' will be addressed later.

13.2   The Calculation of Incremental Resistance

Before the small-signal analysis of a circuit can be carried out, the incremental resistance of each nonlinear component must be found. There are many nonlinear components that could appear in a circuit, so to illustrate the method of analysis we shall assume that the circuit contains one *exponential diode,* so-called in view of the relation it imposes between voltage and current.

The symbol for an exponential diode and the general nature of its current~ voltage relation were shown in Figure 13.1(a). The relation between the diode current i_D and diode voltage v_D can be derived from a knowledge of semiconductor physics and is

$$i_D = I_S[\exp(v_D/v_T) - 1] \tag{13.2}$$

where v_T is called the 'thermal voltage' and has a value of 25 mV at room temperature. When the diode is in its 'forward' bias condition (v_D is positive) the current rises rapidly with increase in voltage: when reverse-biased by more than 100 mV the reverse current is approximately I_S in magnitude. I_S is usually very small and can be as low as 10^{-8}A.

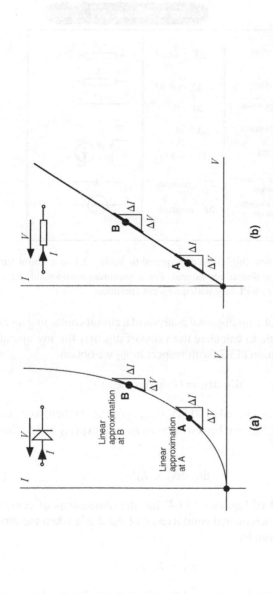

Figure 13.1   (a) The general form of the characteristic of a diode, showing the variation of $\Delta I / \Delta V$ with quiescent condition; (b) a reminder that for a linear resistor the slope $\Delta I / \Delta V$ is constant

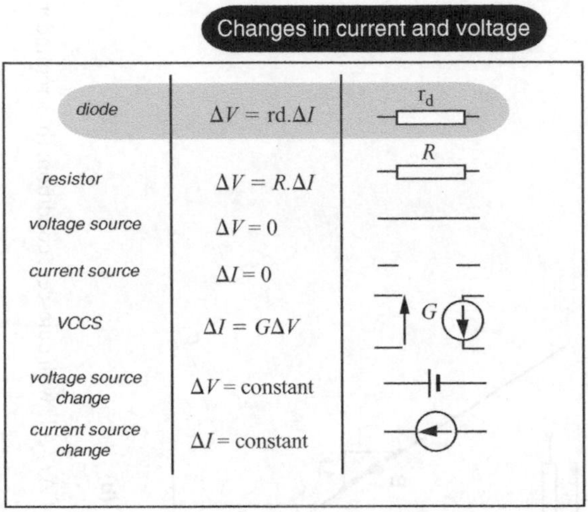

Figure 13.2 Modification (highlighted) required to Table 12.1 to account for the small-signal operation of a nonlinear component. For a nonlinear component the value of the incremental resistance r_d will vary with quiescent condition

In order to carry out a small-signal analysis of a circuit containing an exponential diode we must be able to calculate the value of di_D/dv_D for any operating point. Differentiating Equation (13.2) with respect to v_D we obtain

$$di_D/dv_D = (I_S/v_T)\exp(v_D/v_T) \qquad (13.3)$$

If the diode voltage v_D exceeds 100 mV (i.e., $v_D/v_T \gg 1$) then, to a good approximation, Equation (13.2) can be rewritten as $i_D = I_S\exp(v_D/v_T)$ so that Equation (13.3) becomes

$$di_D/dv_D = I_D/v_T \qquad (13.4)$$

The term on the left of Equation (13.4) has the dimensions of conductance, so we can say that the incremental conductance of the diode when the direct current through it is I_D is given by

$$g_d = I_D/v_T \qquad (13.5)$$

If, as pointed out at the beginning of this chapter, we denote changes in currents and voltages around their average values by lowercase i and v, both with lowercase subscripts, we can rewrite Equation (13.4) as:

$$i_d = g_d\, v_d \qquad (13.6)$$

or, if one's preference is for working in terms of resistance,

$$v_d = r_d \, i_d \tag{13.7}$$

where $r_d = 1/g_d$.

The advantage of the relation shown as Equation (13.6) is, of course, that it is linear. It is also useful to ask whether Eqaution (13.5) appears reasonable. Examination of Figure 13.1(a) will show that it is, because at the lower quiescent condition (A) – i.e., the lower value of I_D – the slope is smaller.

From this, we see that, when applying change analysis to a circuit containing an exponential diode, that diode must be represented in the change model by a resistance having a value r_d equal to the thermal voltage v_T ($= 25$ mV) divided by the quiescent diode current I_D .

How small is 'small'?

We have suggested – and illustrated in Figure 13.1(a) – that it might be acceptable to approximate a limited region of a nonlinear characteristic by a linear segment. How small must this linear segment be for the approximation to be acceptable? The answer can be obtained by taking the expression for diode current (Equation 13.2), by assuming that v_D is sufficiently high for the minus one term to be negligible, and by expressing the diode voltage v_D as the sum of its quiescent and signal components:

$$i_D(t) = I_S \exp[\{V_D + v_d(t)\}/v_T] = I_S \exp(V_D/v_T)\exp(v_d(t)/v_T)$$

Recalling the series expression for e^x:

$$e^x = 1 + x + x^2/2 + x^3/6 + \ldots .$$

we can write that, if $v_d(t)/v_T \ll 1$,

$$i_D(t) = I_S \exp(v_D/v_T)[1 + v_d(t)/v_T]$$

Since, to a good appoximation, $I_S \exp(v_D/v_T) = I_D$, the quiescent value of the diode current, the signal component of $i_D(t)$ is

$$i_d(t) = I_D \, v_d(t)/v_T$$

which is in agreement with Equation (13.6). Thus we see that a linear approximation is valid if $v_d(t) \ll v_T$.

This result has been derived for an exponential diode, and is not necessarily valid for other nonlinear components.

Example 13.1

Figure 13.3 shows a circuit containing one exponential diode. We are required to calculate the total voltage appearing across the diode, i.e., both average and time-varying (signal) component. Because the method of obtaining the solution is common to many problems involving small-signal analysis, we shall formalize the solution into five identifiable steps (*in italics*) and concurrently apply those solution steps to the circuit of Figure 13.3 (non-italic text).

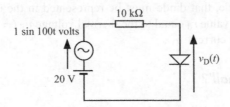

Figure 13.3   A circuit containing a nonlinear component (an exponential diode) and both a constant source and a small-signal source

Step 1

If the circuit contains one or more nonlinear components, find their quiescent conditions with signal amplitudes set to zero.

We shall assume in this example that the direct current through the diode is sufficient to establish a voltage v_D of 0.7 V across the diode. Then, by Ohm's law, the current through the resistor (and hence the diode) is $(20 - 0.7)/10 = 1.93$ mA. This is the value of I_D required in Equation 13.5.

Step 2

Find the small-signal model of each nonlinear component at its quiescent condition.

From Equation (13.5), $g_d = I_D/v_T = 1.93/25 = 0.077$ S, or $r_d = 12.95$ Ω

Step 3

Create the small-signal equivalent of the actual circuit by replacing each component with its small-signal model.

See Figure 13.4

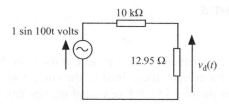

Figure 13.4 The small-signal equivalent of the circuit of Figure 13.3

Step 4

Analyse the small-signal equivalent circuit to find the signal components of the required voltages and currents.

By the voltage divider principle, $v_d(t) = [12.95/(10000 + 12.95)] \times 1 \sin 100t = 1.29 \sin 100t$ mV, to a good approximation

As a reminder, we mention that none of the constant voltages in the actual circuit (e.g., the 20 V source and the 0.7 V across the diode) appear in the small-signal equivalent circuit.

Step 5

Check that signal amplitudes are sufficiently small for the linearity

*assumption to be
valid (e.g., $v_d/v_T \ll 1$
for an exponential diode).*

For the diode, $v_d(t)$ is always much less than 25 mV because its maximum value is 1.29 mV as found in step 4.

The original problem was to find the voltage $v_D(t)$ across the diode. This is the sum of its quiescent and signal components, i.e.:

$$v_D(t) = 0.7 + 0.00129 \sin 100t \text{ V}$$

13.3 Problems

Problem 13.1

Find the incremental resistance of the exponential diode in the circuit of Figure P13.1 for the quiescent current determined by the current source. If the value of the current source is increased by 0.1 mA what change can be expected in the diode voltage v_D?

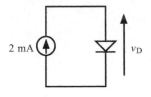

Figure P13.1

Problem 13.2

For the circuit shown in Figure P13.2 determine the quiescent current flowing in the diodes by making a reasonable assumption about the voltage across each diode.

For each diode determine its incremental resistance

Under the assumption that the capacitors possess negligible impedance at the frequency of the sinusoidal voltage source, draw the small-signal equivalent circuit of the actual circuit. Hence calculate the amplitude of the sinusoidal voltages $v_a(t)$ and $v_b(t)$.

If the amplitude of the sinusoidal voltage source is increased above its current value of 2 mV, what is the approximate maximum value it can assume without violating the assumption that both diodes are operating in an essentially linear manner?

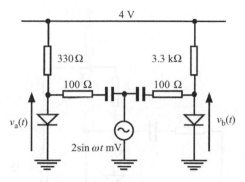

Figure P13.2

Problem 13.3

The circuit shown in Figure P13.3 is a voltage-controlled voltage divider: the direct voltage V determines the quiescent current through the exponential diodes, thereby affecting their incremental resistance. The terms $v_{in}(t)$ and $v_{out}(t)$ denote, respectively, the small-signal voltages at the input and output of the voltage divider.

Assume that the capacitor has negligible impedance at the frequency of the sinusoidal source, and that the voltage V can range between 5 and 20 V.

For each of the extreme values of V (5 and 20 V) calculate the small-signal voltage amplification v_{out}/v_{in} and the maximum magnitude of v_{out} for the calculation to be reasonably accurate. For each diode it may be assumed that, for $i_D > 0.1$ mA, $v_D = 0.7$ V.

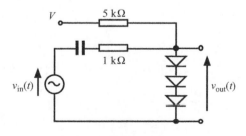

Figure P13.3

Problem 13.4

In the circuit of Figure P13.4 the voltage across each exponential diode can be assumed to be approximately 0.7 V if the diode current exceeds 0.2 mA.

Apply Kirchhoff's current law at point X and hence determine the quiescent voltage at this point and the current in each of the diodes.

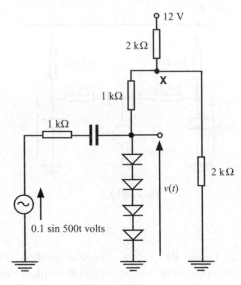

Figure P13.4

Determine the small-signal resistance of each diode.

Assuming that the impedance of the capacitor is negligible, draw the small-signal equivalent of the circuit and hence calculate the peak-to-peak amplitude of the sinusoidal voltage $v(t)$.

Appendix: Answers to Problems

Answer 3.1

(a) $I = 0, V = 6\,\text{V}$

(b) $I = 0, V = -10\,\text{V}$

(c) $I = 2\,\text{mA}$

(d) $I = -4\,\text{mA}, V = 0$

(e) $V = 14\,\text{V}$

(f) $I = 2\,\text{mA}$

(g) $V = -3\,\text{V}$

(h) $I = 4\,\text{mA}, V = -10\,\text{V}$

(i) $I = 6\,\text{mA}$

(j) $V = 8\,\text{V}, V = 4\,\text{V}$

(k) $V = 14\,\text{V}, V = 20\,\text{V}$

(l) $V = -14\,\text{V}$

(m) $V = 2\,\text{V}$

(n) $I = 1\,\text{mA}$

(o) $V = -8\,\text{V}$

Introductory Circuits   Robert Spence
© 2008 John Wiley & Sons, Ltd

(p) $I = 4\,\text{mA}, I = 2\,\text{mA}, V = 8\,\text{V}$

(q) $I = 0, V = 3\,\text{V}$

(r) $V = 2\,\text{V}, I = 26/7\,\text{mA}$

(s) $V = -2\,\text{V}, I = 2\,\text{mA}$

(t) $I = 3\,\text{mA}, V = 0$

Answer 3.5

Equivalence: $6\,\text{k}\Omega$ and $6\,\text{k}\Omega$ in series is equivalent to a $12\,\text{k}\Omega$ resistance.

Equivalence: $12\,\text{k}\Omega$ in parallel with $4\,\text{k}\Omega$ is equivalent to $3\,\text{k}\Omega$.

Equivalence: $3\,\text{k}\Omega$ in series with $4\,\text{k}\Omega$ and $3\,\text{k}\Omega$ is equivalent to $10\,\text{k}\Omega$.

Equivalence: $10\,\text{k}\Omega$ in parallel with the actual $10\,\text{k}\Omega$ resistor is equivalent to $5\,\text{k}\Omega$.

We now have the circuit of Figure A3.5(a).

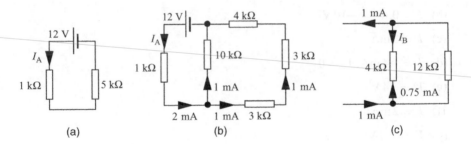

Figure A3.5

From Ohm's law, $I_A = 2\,\text{mA}$.
We now 'back track' to examine the actual $10\,\text{k}\Omega$ in parallel with the equivalent $10\,\text{k}\Omega$ (see Figure A3.5b). As shown, the current of $2\,\text{mA}$ will split equally because the voltage across each $10\,\text{k}\Omega$ resistance is the same.
Now (Figure A3.5c) we examine what the equivalent $3\,\text{k}\Omega$ resistance represents. The current of $1\,\text{mA}$ will split to result in the same voltage across the 4 and $12\,\text{k}\Omega$ resistances, so $I_B = -0.75\,\text{mA}$

Answer 3.7

By Ohm's law $V_1 = -(2\,\text{mA}) \times (1\,\text{k}\Omega) = -2\,\text{V}$.

By KCL, the current through the 'vertical' $1\,\text{k}\Omega$ resistor is $2 + 2 = 4\,\text{mA}$.

By Ohm's law, $V_2 = (4\,\text{mA}) \times (1\,\text{k}\Omega) = 4\,\text{V}$.

Note that V_1 and V_2 are completely independent of the $17\,\text{k}\Omega$ and $2\,\Omega$ resistors.

Answer 3.10

For convenience of reference we redraw the circuit and label the nodes A, B and C (Figure A3.10).

KCL at node C gives $I_2 = 6$ A.

From Ohm's law, the voltage V_{AC} across the 3 Ω resistor is 18 V.

From Ohm's law, the voltage V_{BC} across the 4 Ω resistor is 8 V.

KVL applied around the loop A–B–C–A gives $18 - V - 8 = 0$ so that $V = 10$ V.

KCL at node B indicates a current of 1 A flowing towards node A through the voltage source. Therefore, application of KCL at node A shows that $I_1 = 5$ A.

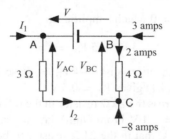

Figure A3.10

Answer 3.11

Device Y takes 1 mA at 2 V and can therefore be represented by a resistance of 2 kΩ. Similarly, device X can be represented by a resistance. We see that the 9 V source is connected to a purely resistive circuit which, by the use of equivalences, could be represented by a single resistor. Current therefore flows out of the positive terminal of the voltage source. That current must be 1.5 mA for the device X to operate correctly. Since the voltage across the device X is 4 V, the voltage V (see

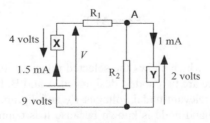

Figure A3.11

Figure A3.11) will (by KVL) be 5 V. Again by KVL the voltage across R_1 is therefore 3 V. But since the current through it is 1.5 mA, Ohm's law tells us that $R_1 = (3\,\text{V})/(1.5\,\text{mA}) = 2\,\text{k}\Omega$. Since device Y takes 1 mA, KCL applied to node A tells us that the current through R_2 is 0.5 mA. Since the voltage across R_2 is 2 V, Ohm's law gives $R_2 = (2\,\text{V})/(0.5\,\text{mA}) = 4\,\text{k}\Omega$.

Answer 4.3

Application of KCL at node A (IN) gives

$$-\frac{V_A}{1} + \frac{(V_B - V_A)}{2} = 0 \text{ which simplifies to}$$

$$-3\,V_A + V_B = 0 \tag{4.1}$$

Application of KCL at node B (IN) gives

$$1 + \frac{(V_A - V_B)}{2} - \frac{V_B}{3} = 0 \text{ which simplifies to}$$

$$3\,V_A - 5\,V_B = -6 \tag{4.2}$$

Adding these two nodal equations gives $-4\,V_B = -6$ so $V_B = 1.5$ V. Substitution in Equation (4.1) gives $V_A = 0.5$ V.

By KVL the voltage across the $2\,\text{k}\Omega$ resistor in the reference direction shown in Figure A4.3 is $V_A - V_B = -1$ V. From Ohm's law we can find the current in the $1\,\text{k}\Omega$ resistor to be 0.5 mA, that in the $2\,\text{k}\Omega$ resistor to be 0.5 mA and that in the $3\,\text{k}\Omega$ resistor to be 0.5 mA, all in the reference directions added to the diagram in Figure 4 A.3. As a check, the currents into nodes A and B do obey KCL.

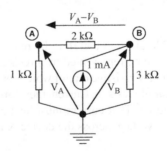

Figure A4.3

Answer 4.8

Refer to Figure A4.8, which shows the selected reference node (indicated by an 'earth' symbol). There are two nodes, designated A and B, for which the voltage is unknown, and the relevant nodal voltages V_A and V_B are shown. The voltage at the extreme right-hand node is known because it is connected directly to the reference node by a voltage source.

We apply KCL at nodes A and B and choose (arbitrarily) to sum currents entering the nodes. Thus, using the 'mA, V, kΩ' set of units,

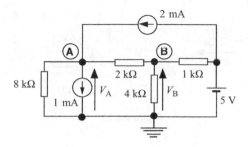

Figure A4.8

KCL at A (IN):

$$2 + (-1) + \frac{(V_B - V_A)}{2} - \frac{V_A}{8} = 0$$

$$\text{giving} - V_A(5/8) + V_B/2 = -1 \tag{4.3}$$

KCL at B (IN):

$$\frac{(V_A - V_B)}{2} + \frac{(5 - V_B)}{1} - \frac{V_B}{4} = 0$$

$$\text{giving } V_A/2 - V_B(7/4) = -5 \tag{4.4}$$

Equations (4.3) and (4.4) are one possible set of nodal equations describing the circuit. There are two linear equations in two unknown voltages, which can therefore easily be solved to find V_A and V_B.

Answer 4.10

The circuit for calculating the voltage across the 12 kΩ resistor due to the voltage source is shown in Figure A4.10(a). By voltage divider action the voltage is seen to be 4 V. Note that this result is independent of the 3 kΩ and unknown resistors because they cannot affect the voltage across a voltage source.

The circuit for calculating the voltage across the 12 kΩ resistor due to the current source is shown in Figure A4.10(b). A current of 1 mA flows through the parallel combination of the 6 and 12 kΩ resistors (equivalent to a 4 kΩ resistor), setting

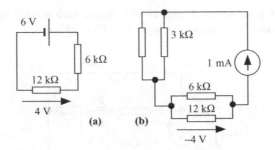

Figure A4.10

up a voltage of –4 V across them in the same reference direction as employed in Figure A4.10(a).

The actual voltage across the 12 kΩ resistor in the actual circuit is, by the superposition principle, the sum of the voltages due to the independent sources taken alone: thus, the voltage is $4 - 4 = 0\,\text{V}$

Answer 4.12

To find the Thevenin equivalent circuit we first find the open-circuit voltage V_{OC} between terminals A and B. By voltage divider action $V_{OC} = 4\,\text{V}$.

We next set the voltage source to zero – in other words, replace it by a short-circuit – and find the resistance between terminals A and B to find the Thevenin resistance R_O. R_O is seen to be the parallel connection of a 6 and a 4 kΩ resistor, and equivalent to 2.4 kΩ, The Thevenin equivalent circuit is shown in Figure A4.12(a).

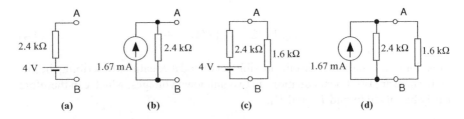

Figure A4.12

We can find the Norton equivalent of the original circuit directly from the Thevenin equivalent. The current source (Figure A4.12 b) is V_{OC} divided by R_O, giving $I_{SC} = 1.667\,\text{mA}$, and the resistance in the Norton equivalent is the same as for the Thevenin equivalent.

We now connect a 1.6 kΩ resistor between terminals A and B using both the equivalent circuits (Figure A4.12 c and A4.12 d). When using the Thevenin equivalent circuit the current through the $1.6\,k\Omega$ resistor is easily found by Ohm's law as $4/(2.4 + 1.6) = 1\,mA$.

Just as the voltage divider principle can be applied to a circuit of the form of Figure A4.12(c), a 'current divider' principle can be applied to analyse the circuit of Figure A4.12(d). The conductance of the $1.6\,k\Omega$ resistor is $0.625\,mS$, and that of the $2.4\,k\Omega$ resistor is $0.4167\,mS$. By the current divider principle the current through the $1.6\,k\Omega$ resistor is $(1.667).[0.625/(0.625 + 0.4167)]$ which is $1\,mA$.

Answer 4.13

The open-circuit voltage between terminals A and B is, by KVL, $-8 + (5\,k\Omega \times 4\,mA) = 12\,V$ (there is no voltage drop across the right-hand $5\,k\Omega$ resistor because no current flows in or out of terminal A).

To find the Thevenin resistance R_O we set the values of the voltage source and current source to zero, thereby resulting in two $5\,k\Omega$ resistors connected in series between A and B: the equivalent resistance, R_O, is $10\,k\Omega$.

The connection of a $12\,V$ source in the polarity described would result in no current flow (since the external $12\,V$ source directly opposes the internal $12\,V$ source in the Thevenin model).

Answer 5.2

We select a voltage reference node indicated by the earth symbol in Figure A5.2 and label the nodes (A, B) for which the nodal voltage is unknown.

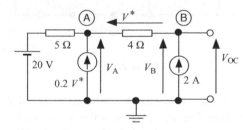

Figure A5.2

By application of KCL at node A we obtain:

$$\frac{(20 - V_A)}{5} + \frac{(V_B - V_A)}{4} + 0.2(V_A - V_B) = 0$$

which can be rearranged as

$$-0.25\,V_A + 0.05\,V_B = -4 \qquad (5.1)$$

The application of KCL at node B leads to

$$2 + \frac{(V_A - V_B)}{4} = 0$$

which can be rearranged as

$$0.25\,V_A - 0.25\,V_B = -2 \qquad\qquad (5.2)$$

Addition of these two nodal Equations (5.1) and (5.2) yields $V_B = 30\,\text{V}$. Since V_B and V_{OC} are identical, $V_{OC} = 30\,\text{V}$.

Answer 5.4

To find the resistance between the external terminals of the grey box we connect (Figure A5.4) a current of 1 A between them and calculate the voltage V that appears: the ratio $V/(1\,\text{A})$ is then the required resistance.

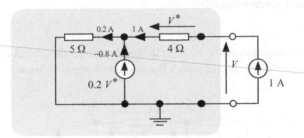

Figure A5.4

From the circuit we see, from Ohm's law, that $V^* = -4\,\text{V}$. The value of the controlled source is therefore $0.2\,(-4) = -0.8\,\text{A}$. By KCL it follows that the current through the $5\,\Omega$ resistor is 0.2 A (right to left). By Ohm's law the voltage across the $5\,\Omega$ resistor is 1 V. Using KVL we can write

$$V = 1 + 4 = 5\,\text{V}.$$

A two-terminal box whose current is 1 A and whose voltage is 5 V is equivalent to a resistance R_O of $5\,\Omega$.

Answer 5.5

We first create the circuit of Figure A5.5(a) to calculate the current I due to the voltage source acting alone. With the reference node for voltage as shown we apply KCL at node A to obtain

$$-\frac{V}{1} + 0.2\,V + \frac{(3-V)}{1} = 0$$

which gives $V = 1.666$ V. By KVL the voltage across the top $1\,k\Omega$ resistor is 1.333 V and the current $I = -1.333$ mA in the reference direction shown.

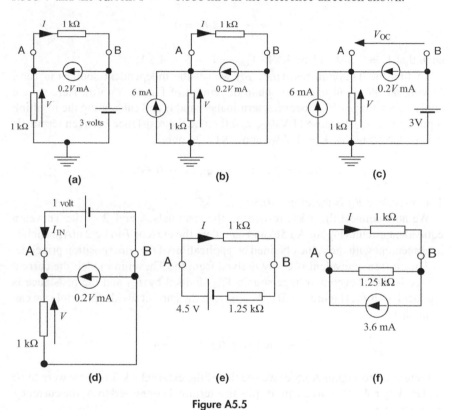

Figure A5.5

We now create the circuit of Figure A5.5(b) to calculate the current I due to the independent current source acting alone. With the reference node as shown, application of KCL at node A gives

$$6 - \frac{V}{1} + 0.2\,V - \frac{V}{1} = 0$$

which gives $V = 3.333$ V and, by Ohm's law, $I = 3.333$ mA in the reference direction shown.

Application of the superposition principle allows us to say that the actual value of I in the original circuit in the reference direction shown is $3.333 - 1.333 = 2\,\text{mA}$,

To obtain the requested Thevenin equivalent circuit we first find the open-circuit voltage V_{OC} between terminals A and B (see Figure A5.5c). With the chosen voltage reference node, application of KCL at node A leads to

$$6 - \frac{V}{1} + 0.2\,V = 0$$

such that $V = 7.5\,\text{V}$ and, by KVL, $V_{OC} = V - 3 = 4.5\,\text{V}$.

To find the Thevenin resistance R_O we set the independent sources to zero, *leaving the VCCS in place*, to obtain the circuit of Figure A5.5(d). Here we are applying a voltage of 1 V between terminals A and B and calculating the resulting input current I_{IN}, because $(1\,\text{V})/I_{IN}$ will then be the resistance between terminals A and B. Noting that $V = 1\,\text{V}$ we apply KCL to obtain

$$I_{IN} + 0.2 = 1/1 \text{ giving } I_{IN} = 0.8A.$$

The resistance R_O is therefore $1/0.8 = 1.25\,\text{k}\Omega$.

We now connect the 1 kΩ resistor to the terminals A and B of the Thevenin equivalent circuit (Figure A5.5e) and calculate the current I to be 2 mA, which is in agreement with the value obtained by application of the superposition principle.

The Norton equivalent is easily derived from the Thevenin model: the current source is the Thevenin voltage source V_{OC} divided by R_O and the resistance is identical with R_O (Figure A5.5f). By applying the current divider principle we can write that

$$I = 3.6[1/1 + 0.8)] = 2 \text{ mA}.$$

By reference to Figure A5.5(e) we see that if the external 1 kΩ resistor were to be replaced by a 4.5 V source with its positive terminal connected to A, the current I would be zero.

Answer 5.6

To find the Thevenin equivalent circuit we first analyse the circuit with nothing attached to the external terminals (Figure A5.6a). Application of KCL at node A leads to

$$\frac{(20 - V_{OC})}{1} - (V_{OC} - 20) - \frac{V_{OC}}{1} = 0 \text{ which gives}$$

$V_{OC} = 13.33\,\text{V}.$

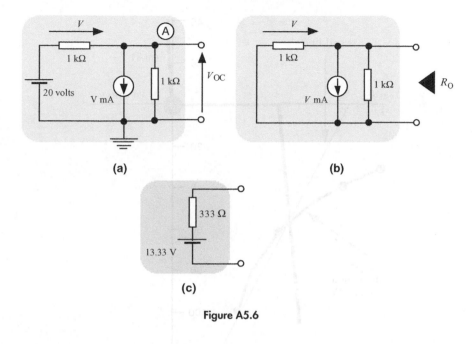

(a)

(b)

(c)

Figure A5.6

To find the Thevenin resistance R_O we set the independent voltage source equal to zero and remember not to set the dependent source to zero (Figure A5.6b). The calculation of the resistance R_O between the external terminals is considerably simplified by observing that the voltage controlling the current source appears directly across the current source, which is therefore equivalent to a resistance, in this case of value $1\,k\Omega$ (A voltage V across a two-terminal component creating a current of $V\,mA$ through it is characteristic of a resistance of value $1\,k\Omega$). Thus, by reference to the figure, we have three $1\,k\Omega$ resistors in parallel, equivalent to 333Ω. Thus, $R_O = 330\,\Omega$. The complete Thevenin equivalent circuit is shown in Figure A5.6(c).

Answer 5.9

In Figure A5.9 the added points correspond to a constant product (300 mW) of V and I, and the boundary sketched in indicates the permissible region of operation. The plotted load-line is associated with the series connection of the 8 V source and the resistor R for the case in which the intersection of the load-line and the Zener diode characteristic is at the minimum, for R, consistent with the 300 mW limit of dissipated power. If the intersection of the load-line with the diode current axis is estimated to be 160 mA, the corresponding value of R is $(8\ V)/(160\ mA) = 50\ \Omega$.

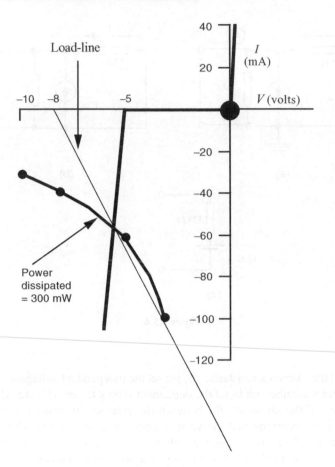

Figure A5.9

The load-line intersects the Zener characteristic at about 5.5 V, so the voltage across the resistor R would be $8 - 5.5 = 2.5$ V. The intersection indicates a current of approximately 50 mA, so the power dissipated by the resistor is $(2.5 \text{ V}) \times (50 \text{ mA}) = 125$ mW.

Answer 6.1

(a) $V^+ > V^-$ therefore $V_I > 0$. Hence V_O is also positive and equal to $+10$ V.

(b) $V^+ = 0$, so $V_I = V^+ - V^- = 0 - 4 = -4$ V. Since V_I is negative V_O must be negative and equal to -10 V.

(c) $V_1 = (-5) - (-4) = -1$ V. Since V_1 is negative V_O must be equal to -10 V.

(d) $V_1 = 4 - 5 = -1$ V. Therefore $V_O = -10$ V.

Answer 6.3

The circuit is a cascade of two triggers.

We first examine the trigger involving opamp X in order to determine the threshold values at which the voltage at A will cause a change of state. When the voltage at C is 15 V, then $V^+ = 4$ V. Therefore the voltage at C changes from $+15$ to -15 V when the voltage at C exceeds 4 V. Similarly, the voltage at C changes from -15 to $+15$ V when the voltage at A falls below -4 V.

We now examine the trigger involving opamp Y. To establish the threshold voltage levels pertinent to the voltage at C we examine the circuit of Figure A6.3(a). The opamp Y changes state (voltage at B changes from $+15$ to -15 V) when the voltage at X just starts to go negative. With $V_X = 0$ the current through the $10\,\text{k}\Omega$ resistor is 1.5 mA. This current flows through the $5\,\text{k}\Omega$ resistor, creating a voltage of 7.5 V so that the voltage at C is -7.5 V. Thus the output of opamp Y changes from $+15$ to -15 V when the voltage at C decreases below -7.5 V, and it changes back from -15 to $+15$ V when the voltage at C increases above 7.5 V.

At $t = 2\,T$ the voltage at A increases above 4 V, so the voltage at C drops from $+15$ to -15 V. As a result the voltage at B also drops to -15 V.

At $t = 4\,T$ there is no effect because the threshold for the voltage at A is now -4 V.

At $t = 4.5\,T$ the voltage at A decreases below -4 V so the voltage at C changes from -15 to $+15$ V. As a result the voltage at B also changes from -15 to $+15$ V.

Waveforms of the voltages at B and C are shown in Figure A6.3(b)

The current I has two components. The current through the series connection of the 11 and $4\,\text{k}\Omega$ resistors (equivalent to $15\,\text{k}\Omega$ since no current is drawn by the positive input terminal of opamp X) is 1 mA when the voltage at C is $+15$ V and -1 mA when it is -15 V. The other component of the current I is the current through the $5\,\text{k}\Omega$ resistor. But since the voltages at C and B are always identical this current has a value of zero. Thus, the waveform of the current I is as shown in Figure A6.3(b).

Refer to the circuit of Figure A6.3c. If the voltage at B stays at 15 V when the voltage at C falls to -15 V the voltage V^+ at node X is -5 V because the current through the $10\,\text{k}\Omega$ resistor is $(30\,\text{V})/15\,\text{k}\Omega = 2\,\text{mA}$. To maintain the input voltage V_1 of opamp Y positive (to keep B at $+15$ V) the voltage V^- at the negative input terminal of opamp Y must therefore be less than -5 V. The connection of a 6 V source (note the polarity shown in Figure A6.3 c) would ensure that the value of V_1 for opamp Y is positive so that the voltage at B remains unchanged at $+15$ V.

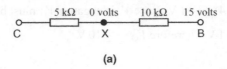

(a)

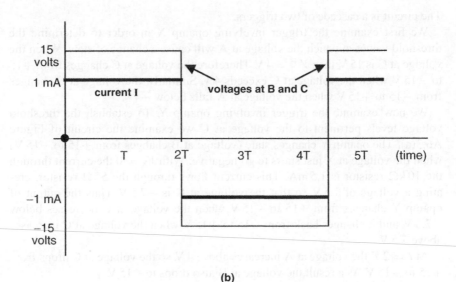

(b)

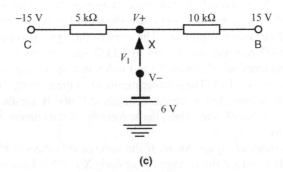

(c)

Figure A6.3

Answer 6.4

The same voltage V is applied to two comparators (but note the polarity of the opamp input terminals). For any value of V, therefore, it is possible to determine the output voltage of each opamp. Only when these two voltages are the same will no current flow in the $10\,k\Omega$ resistor.

Corresponding values of V and the outputs V_L and V_R of the left-hand and right-hand opamps respectively are (in volts):

V	-1	0	0.99	1.1	1.9	2.1	3
V_L	-10	-10	-10	-10	-10	$+10$	$+10$
V_R	$+10$	$+10$	$+10$	-10	-10	-10	-10

In fact, the critical transitions occur when $V = 1.0$ and 2.0 V. Within that range the output voltages of the two opamps are identical and no current will flow through the $10\,k\Omega$ resistor.

Answer 7.3

Since no current enters the negative input terminal of the opamp we can apply the voltage divider principle to calculate the voltage at the junction of the 4 and $1\,k\Omega$ resistors: it is $(4/5) \times V = 0.8V$. Again, because there is no voltage drop across the $11\,k\Omega$ resistor, the voltage at the negative input terminal is also $0.8V$.

With negative feedback we assume a virtual short circuit between the input terminals of the opamp, so the voltage at the positive input terminal is also $0.8V$. The voltage at this point is also equal to 4 V because there is no current through, and therefore no voltage across, the $17\,k\Omega$ resistor. Thus,

$$0.8\,V = 4 \text{ giving } V = 5\,V.$$

Note the redundant nature of the 17 and $11\,k\Omega$ resistors.

Answer 7.5

Since the voltage at the negative input terminal of the opamp is essentially the same as at the positive input terminal (i.e., there is a virtual short-circuit between the two terminals) and hence at earth voltage, the direct voltage provided by the voltage source appears directly across the photodiode and provides it with the reverse bias necessary to its successful operation. The only factor that controls the current through the photodiode is the incident radiation.

Any current I_D passed by the diode must flow through the resistor R, setting up a voltage $I_D R$ across it. Thus, the voltage measured by the voltmeter is, by KVL, equal to $-I_D R$ since the negative input terminal is essentially at zero voltage. For each microwatt of radiation the diode generates $0.5\mu A$ and sets up a measured voltage having a magnitude of $(0.5\mu A) \times R$. But we are told that the scale factor of the voltmeter is to be 2.5 µW/mV, so the measured voltage corresponding to a microwatt of incident radiation is 0.4 mV. Thus,

$$(0.5\,\mu A) \times R \times (2.5) = 1\,mV, \text{ giving } R = 800\,\Omega.$$

Answer 7.7

Because there is a virtual short-circuit between the input terminals of each opamp, the voltage V_2 appears at the top of the $10\,k\Omega$ resistor and the voltage V_1 at the bottom. The voltage across the $10\,k\Omega$ resistor is therefore $(V_1 - V_2)$ and the current through it is $(V_1 - V_2)/10\,k\Omega$ upwards. All this current must flow in the $50\,k\Omega$ resistors because no current can flow into the negative terminals of the opamps. Thus, by Ohm's law,

$$V_Y = V_2 - 5(V_1 - V_2) = -5\,V_1 + 6\,V_2$$
$$V_X = V_1 + 5(V_1 - V_2) = 6\,V_1 - 5\,V_2$$

In calculating V_{OUT} we can regard V_X and V_Y as fixed at the values given above and we can otherwise ignore the circuit to the left of the $5\,k\Omega$ resistors.

We now employ superposition to calculate the effects of V_X and V_Y. If we set $V_X = 0$ the voltage at the positive input terminal of the right-hand opamp is zero. Recalling the expression for the voltage gain of an inverter we can write:

$$V_{OUT} \text{ due to } V_Y = -(50\,k\Omega/5\,k\Omega) \times V_Y = -10\,V_Y$$

If we now set $V_Y = 0$ the voltage at the negative input terminal of the right-hand opamp is, by voltage divider action, $V_{OUT}[5/(5+50)] = V_{OUT}/11$. This must also be the voltage at the positive input terminal in view of the virtual short-circuit between the input terminals. But, again by voltage divider action, this voltage must be $V_X[50/(50+5)]$. Thus,

$$V_X(50/55) = V_{OUT}/11 \text{ giving } V_{OUT} \text{ (due to } V_X) = 10\,V_1$$

Adding the contributions of V_X and V_Y we find that

$$V_{OUT} = 10(V_X - V_Y) = 110(V_1 - V_2).$$

Answer 7.9

First, we calculate the open-circuit voltage between terminals A and B.

With negative feedback applied to both opamps, a virtual short-circuit will occur between their input terminals (see Figure A7.9a). Because $V_1 = 0$ for each opamp, the voltage of 6 V appears across the 6 kΩ resistor, giving rise to a current of 1 mA flowing in the direction X to Y. That current must flow through the 10 and 1 kΩ resistors, setting up voltages of 10 and 1 V, respectively. So, the voltage between P and Q is, by Ohm's law, 1 mA$\times$(10 kΩ + 6 kΩ + 1 kΩ) = 17 V. There is no current through the 5 kΩ resistor and therefore no voltage across it, so the open-circuit voltage is 17 V.

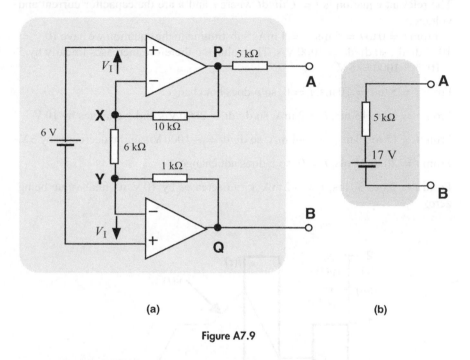

(a)   (b)

Figure A7.9

To find the Thevenin resistance R_O we set the independent sources to zero: in this case we replace the 6 V source by a short circuit. There is now no voltage across the 6 kΩ resistor and hence no current through it. There is therefore no current through the 10 and 1 kΩ resistors. By Ohm's law there is therefore no voltage between terminals P and Q. The resistance between terminals A and B is therefore 5 kΩ.

The Thevenin equivalent circuit of the shaded box is as shown in Figure A7.9(b).

Answer 7.12

No current flows into the positive input terminal of the opamp so we can apply the voltage divider principle to calculate the voltage at that terminal to be zero.

With a virtual short-circuit between the opamp's input terminals the voltage at the negative input terminal must also be zero. The current through the left-hand $4\,\text{k}\Omega$ resistor is therefore, by Ohm's law, 1 mA. Again by Ohm's law, the current through the $8\,\text{k}\Omega$ resistor is zero. Applying KCL at the negative input terminal shows that 1 mA flows through the left-hand $2\,\text{k}\Omega$ resistor, creating a voltage of 2 V across it. Since the voltage at the negative input terminal is zero, $V = -2\,\text{V}$.

Answer 8.1

The relevant equation is $i = C\,dv/dt$ where i and v are the capacitor current and voltage.

From $t = 0$ to $t = 5\,\text{ms}$, $i = 1\,\text{mA}$. Substituting in the equation we have $10^{-3} = 10^{-6}\,dv/dt$, so $dv/dt = 1000\,\text{V/s}$. The voltage v therefore increases linearly by $5 \times 10^{-3} \times 1000 = 5\,\text{V}$.

From $t = 5$ to $t = 10\,\text{ms}$, $i = 0$, so v does not change.

From $t = 10$ to $15\,\text{ms}$, $i = 2\,\text{mA}$, so $dv/dt = 2000\,\text{V/s}$ and v changes by 10 V.

From $t = 15$ to $20\,\text{ms}$, $i = -1\,\text{mA}$, so $dv/dt = -1000\,\text{V/s}$ and v decreases by 5 V.

From $t = 20$ to $25\,\text{ms}$, $i = 0$, so v does not change.

From $t = 25$ to $30\,\text{ms}$, $i = -2\,\text{mA}$ so v decreases by 10 V, its final value being zero.

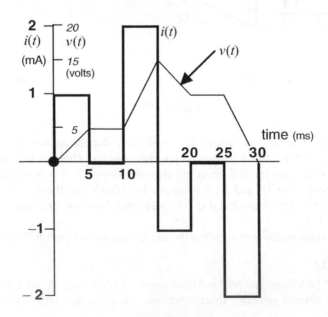

Figure A8.1

Note that this final value is consistent with the fact that the area under the plot of $i(t)$ is zero. Figure A8.1 shows the waveform of $v(t)$ superimposed on the waveform of $i(t)$.

Answer 8.3

The sub-circuit governing the switching of the trigger circuit is shown in Figure A8.3(a), from which we calculate that $R_1 = 8\,\text{V}/1\,\text{mA} = 8\,\text{k}\Omega$.

Current into capacitor when voltage at A is 10 V is $10/R_2$. Substituting in the equation $i = C\,dv/dt$, and observing that the capacitor voltage is equal to $-V_O$ (due to the virtual short circuit between the opamp input terminals) we find that $10/R_2 = 10^{-6} \times 200$ and $R_2 = 50\,\text{k}\Omega$.

A dimensioned sketch of the waveforms $v_O(t)$ and $v(t)$ appears in Figure A8.3(b).

The relation between $v_B(t)$ and $v_O(t)$ for the circuit shown in Figure P8.3(d) is shown in Figure A8.3(c).

When the circuits in Figure P8.3(c, d) are connected as described, the waveform of $v_B(t)$ is as shown dashed in Figure A8.3(b). Note that when $v_O > -4\,\text{V}$, $v_B = 10\,\text{V}$, and when $v_O < -4\,\text{V}$, $v_B = -10\,\text{V}$.

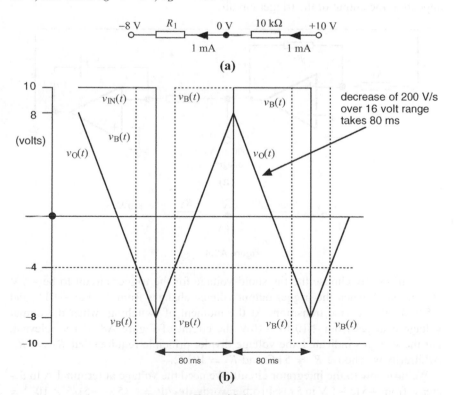

Figure A8.3

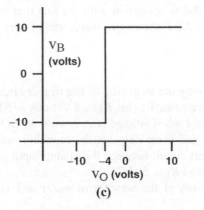

Figure A8.3   (*Continued*)

Answer 8.4

Figure A8.4(a) shows the form of the required circuit, for which the values of the components R_1, R_2, R_3 and C must be calculated. The square-wave voltage appears at the output of the trigger circuit.

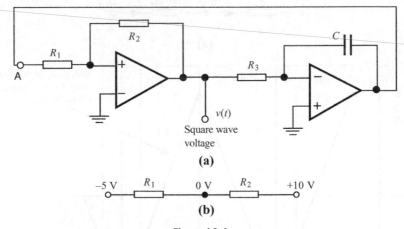

Figure A8.4

We arbitrarily choose the threshold voltage for the trigger circuit to be -5 V at terminal A when the trigger output voltage changes from $+10$ to -10 V, and $+5$ V for the reverse transition. At the moment of switching, when the output voltage changes from $+10$ to -10 V the circuit of Figure A8.4(b) is relevant. For the voltages indicated, the voltage divider principle requires that $R_2 = 2R_1$. Arbitrarily we choose $R_1 = 5$ kΩ and $R_2 = 10$ kΩ.

We turn now to the integrator circuit. We need the voltage at terminal A to decrease from $+5$ to -5 V in 5 ms. In other words, $dv_A/dt = -(5 - -5)/5 \times 10^{-3} =$

-2000 V/s. The current flowing into the capacitor is $(10\,\text{V})/R_3$. Substituting in the equation relating capacitor current and voltage (and recalling that the capacitor voltage is $-v_A$) we can write that

$10/R_3 = C \times 2000$ so that $CR_3 = 5 \times 10^{-3}$.

Arbitrarily we choose $R_3 = 5\,\text{k}\Omega$ and $C = 1\,\mu\text{F}$.

Answer 8.5

Refer to Figure P8.4, and recall that R_2 was chosen to be $50\,\text{k}\Omega$. The connection of the $100\,\text{k}\Omega$ resistor and 10 V source will not affect the threshold voltages for the trigger (they are determined by R_1 and the $10\,\text{k}\Omega$ resistor) or the voltages (±10 V) between which the output voltage of the trigger varies. Thus, the current through R_2 ($50\,\text{k}\Omega$) will continue to be $10/R_2 = 10/50 = 0.2\,\text{mA}$ when $V_{IN} = 10$ V and $-0.2\,\text{mA}$ when $V_{IN} = -10$ V. Before the connection of the $100\,\text{k}\Omega$ resistor and 10 V source these currents alone charged the capacitor. However, because there is a virtual short-circuit at the input to the right-hand opamp, a *constant* current of $10/100 = 0.1\,\text{mA}$ will be added to the charging current. Thus, when $V_{IN} = 10$ V, the capacitor charging current is $0.2 + 0.1 = 0.3\,\text{mA}$, but is $-0.2 + 0.1 = -0.1\,\text{mA}$ when $V_{IN} = -10$ V. From the equation $i = C\,dv/dt$ it follows that the corresponding rates of change of the voltage at the output of the integrator are, respectively, $-0.3/C$ and $0.1/C$. Since $C = 1\,\mu\text{F}$, these rates are $-0.3 \times 10^{-3} \times$

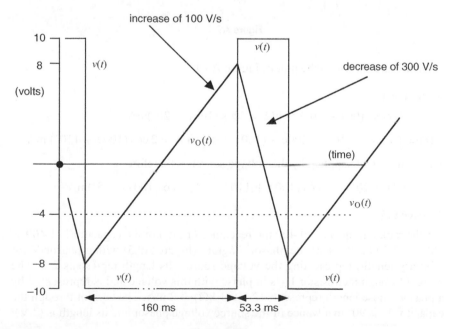

Figure A8.5

$10^6 = -300$ V/s and $0.1 \times 10^{-3} \times 10^6 = 100$ V/s. The corresponding times taken to traverse 16 V are, respectively, 53.3 and 160 ms. The new periodic time of the voltage waveform appearing at A will therefore be $160 + 53.3 = 213.3$ ms. The waveforms of the voltages $v_S(t)$ and $v(t)$ are shown in Figure A8.5.

Answer 9.1

$i(t) = Cd(2\cos 100\,t)/dt = -200\,C\sin 100\,t = -200 \times 10^{-6}$ $(\cos 100\,t + \pi/2)$
$A = 0.2\cos(100\,t + \pi/2)$ mA

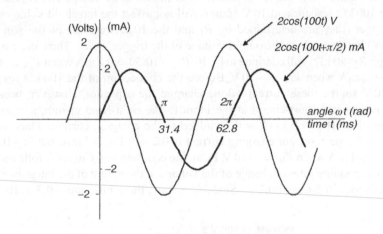

Figure A9.1

Details are shown on the plot in Figure A9.1

Answer 9.4

$i_R(t) = [2\cos(100\,t + 20°)]/1$ kΩ $= 2\cos(100\,t + 20°)$mA

$i_C(t) = Cdv/dt = 10^{-5} \times 200\cos(100\,t + 110°)$ A $= 2\cos(100\,t + 110°)$ mA

The total current supplied by the voltage source is therefore

$2\cos(100\,t + 20°) + 2\cos(100\,t + 110°) = 2\sqrt{2}\cos(100\,t + 65°)$mA

Answer 9.5

For the radian frequency of 400 the reactance $(1/\omega C)$ of the capacitor is $1/400 \times 10^{-6} = 2.5$ kΩ. We start the phasor diagram (Figure A9.5) with a phasor **V** (of arbitrary length) representing the voltage source. Its length represents 2 V. The current through the resistor I_R is in phase with this voltage and is represented by a phasor whose length represents $(2\text{ V})/(1\text{ k}\Omega) = 2$ mA. The current through the capacitor I_C is 90° in advance of the source voltage phasor and its length is (2 V)/

$(2.5 \text{ k}\Omega) = 0.8$ mA. To find the phasor **I** representing the current supplied by the source we carry out the phasor addition of $\mathbf{I_R}$ and $\mathbf{I_C}$ (see Figure A9.5). The length of that phasor represents a current whose magnitude is $\sqrt{(2^2 + 0.8^2)} = 2.15$ mA. Note: otherwise overlapping phasors have been offset slightly for ease of interpretation.

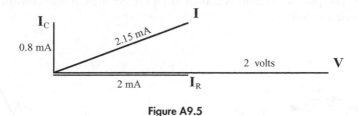

Figure A9.5

Answer 9.7

Refer to the phasor diagram and labelled circuit in Figure A9.7. The reactance of each inductor ωL is 1000 Ω. Assume the phasor **V** is at zero phase angle and has a magnitude V. The current phasor $\mathbf{I_R}$ is therefore of length $V/1000 = V$ mA, also at zero phase angle. The current phasor $\mathbf{I_1}$ lags **V** by 90° and has a magnitude $V/1000 = V$ mA. The current phasor $\mathbf{I_2}$ is, by KCL, the phasor sum of $\mathbf{I_R}$ and $\mathbf{I_1}$: its magnitude is $V\sqrt{2}$ mA and its phase is minus 45° with respect to the phasor **V**. The current represented by $\mathbf{I_2}$ flows through the left-hand inductor, setting up a voltage $\mathbf{V_2}$ which leads $\mathbf{I_2}$ by 90°. The magnitude of $\mathbf{V_2}$ is therefore ($V\sqrt{2}$ mA).(1000 Ω) = $V\sqrt{2}$. The source voltage $\mathbf{V_S}$, whose amplitude we are told is 2 V, is, by KVL, the phasor sum of **V** and $\mathbf{V_2}$ and therefore has a magnitude $V\sqrt{5}$. Since $V\sqrt{5} = 2$, the magnitude of **V** is 0.893 V. Again, in the phasor diagram, overlapping phasors have been offset slightly for ease of interpretation.

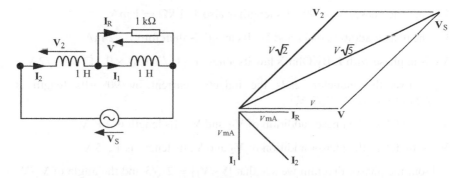

Figure A9.7

Answer 9.8

We first calculate the reactances of the capacitor and inductor. As $\omega = 1000$ the capacitor reactance $(1/\omega C)$ is 1 kΩ. The reactance of the inductor (ωL) is also 1 kΩ.

The circuit is reproduced in Figure A9.8(a) in order to identify voltages and currents. The phasor diagram is shown in Figure A9.8(b). It was constructed in the following sequence:

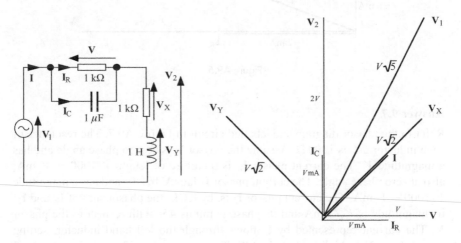

Figure A9.8

V is arbitrarily drawn as shown, of length V.

I_R is in phase with **V**. By Ohm's law its length is $V/1$ k$\Omega = V$ mA.

I_C leads the phasor **V** by 90°. Its length is also $V/1$ k$\Omega = V$ mA.

I is the phasor addition of I_R and I_C. Its length is therefore $V\sqrt{2}$ mA.

V_X is in phase with **I**. By Ohm's law its length is $(V\sqrt{2}$ mA$) \times 1$ k$\Omega = V\sqrt{2}$ V.

V_Y across the inductor leads the inductor current by 90°. Its length is $(V\sqrt{2}$ mA$) \times \omega L = V\sqrt{2}$ V.

V_2 is, by KVL, the phasor addition of V_X and V_Y. Its length is $2 V$ V.

V_1 is, by KVL, the phasor addition of V_2 and **V**. Its length is $V\sqrt{5}$ V.

From the phasor diagram we see that $|V_2/V_1| = 2/\sqrt{5}$ and the angle of V_2/V_1 is $\tan^{-1} 0.5$.

Answer 10.1

A selection:

(a) $Z = j\omega L = j \times 10^3 \times 1 = j$ kΩ, $Y = 1/Z = -j$ mS, $|Z| = 1$ kΩ, $\angle Z = 90°$, $|Y| = 1$ mS, $\angle Y = -90°$

(b) $Z = R = 1$ kΩ, $Y = 1/Z = 1$ mS, $|Z| = 1$ kΩ, $\angle Z = 0$, $|Y| = 1$ mS, $\angle Y = 0$

(c) $Z = 1/j\omega C = 1/j \times 10^3 \times 10^{-6} = -j$ kΩ, $Y = 1/Z = 1/(-j) = j$ mS, $|Z| = 1$ kΩ, $\angle Z = -90°$, $|Y| = 1$ mS, $\angle Y = 90°$

(d) $\omega = 2\pi \times 159 = 1000$ rad/s, $Z = R + j\omega L = 100 + j \times 10^3 \times 10^{-1} = 100 + j100$ Ω, $Y = 1/Z = 1/(100 + j100) = 5 - j5$ mS, $|Z| = 100\sqrt{2}$ Ω, $\angle Z = 45°$, $|Y| = 5\sqrt{2}$ mS, $\angle Y = -45°$

(e) $Y = G + j\omega C = 0.001 + j500 \times 10^{-6} = 1 + j0.5$ mS, $\quad$ $Z = 1/Y = 1/(1 + j0.5) = 0.8 - j0.4$ kΩ, $|Y| = \sqrt{(1 + 0.25)} = 1.12$ mS, $\angle Y \tan^{-1} 0.5 = -26°33'$ $|Z = 0.896$ kΩ, $\angle Z = \tan^{-1} 0.5 = 26°33'$

(f) $Y = j\omega C + 1/j\omega L = j \times 10^3 \times 0.1 \times 10^{-6} + 1/j \times 10^3 \times 0.1 =$ (approx.) $j10^{-2}$ S $= j10$ mS, $Z = 1/Y = -j0.1$ kΩ, $|Y| =$ (approx.) 10 mS, $\angle Y = -90°$, $\angle Z = 90°$, $|Z| =$ (approx.) 0.1 kΩ

Answer 10.2

$I = V_S/(R + 1/j\omega C) = (0 - j10)/[1 + 10^{-3}/(j \times 10^3 \times 10^{-6})] = -j10/(1 - j) = 5 - j5$ mA

$V_R = RI = 5 - j5$ V, $V_C = I/j\omega C = (5 - j5) \times 10^{-3}/j \times 10^3 \times 10^{-6} = -5 - j5$ V. As a check we note that the sum of V_R and V_C is $-j10$ V

If V_S is changed to $10 + j0$ we note that its magnitude is unchanged, but its angle increased by $90°$. The same will then be the case for all other voltages and currents in the circuit. In other words, $I = 5 + j5$ mA, $V_R = 5 + j5$ V and $V_C = 5 - j5$ V (as a check, $V_R + V_C = V_S$)

Answer 10.4

For the current $i(t)$ to be zero irrespective of the amplitude of the voltage source the impedance presented by the two components connected in parallel must be infinite. In other words the admittance must be zero. The admittance of the two components connected in parallel is

$Y = j\omega C + 1/j\omega L = j\omega \times 10^{-6} - j/\omega \times 10^{-2}$.
For $Y = 0$, $\omega = 10^4$ rad/s.

Answer 10.9

See Figure A10.9 in which various complex voltages and currents are identified.

By the voltage divider principle

$$\mathbf{V}_A = [R/(R + j\omega L)]\mathbf{V}_S$$

and

$$\mathbf{V}_B = [(1/j\omega C)/(R + 1/j\omega C)]\mathbf{V}_S$$

By KVL $\mathbf{V} = \mathbf{V}_B - \mathbf{V}_A$

So $\mathbf{V} = \mathbf{V}_S[(1/(1 + j\omega CR) - 1/(1 + j\omega L/R)]$

Substitution of $R = \sqrt{(L/C)}$ shows that $\mathbf{V} = 0$, independent of frequency.

Current $\mathbf{I}_1$ in upper branch is given by $\mathbf{I}_1 = \mathbf{V}_S/(R + j\omega L)$

Current $\mathbf{I}_2$ in lower branch is given by $\mathbf{I}_2 = \mathbf{V}_S/(R + 1/j\omega C)$

The total current supplied by the voltage source is, from KCL, the sum of $\mathbf{I}_1$ and $\mathbf{I}_2$. Substituting $R = \sqrt{(L/C)}$ we find

$$\mathbf{I}_S = \mathbf{V}_S/R$$

In other words, the current supplied by the source is the same as would be supplied if the source were connected to a resistance of value R.

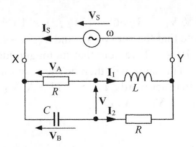

Figure A10.9

Answer 10.12

Box 1
There is no indication of any reactive component since the impedance has no imaginary part. The box could contain one $1\,k\Omega$ resistor, or two resistors in series whose total resistance is $1\,k\Omega$, or two in parallel with a combined conductance of $1\,mS$.

Box 2

An admittance with a real part which is independent of frequency indicates a resistor in parallel with a reactive element. The constant real part is a conductance of 0.001 S, in other words a resistance of 1 kΩ. The imaginary part is seen to be proportional to frequency, indicating a capacitor in parallel with the resistor. The admittance of a capacitor is $j\omega C$ and is seen to be 0.001 S at $\omega = 10^3$. Therefore, $C = 0.001/10^3 = 10^{-6}$ F or 1 μF.

Box 3

The impedance has no real part at either frequency, and its imaginary part is directly proportional to frequency, indicating an inductor. Taking the higher frequency of 10^6 we see that $\omega L = 10^6$ Ω, giving $L = 1$ H.

Box 4

The measurements indicate a series resonant circuit (capacitance and inductance in series), with resonance occurring at $\omega = 10^3$. At the higher frequency, far removed from resonance ($\omega = 10^6$), the impedance will be almost entirely due to the inductance and equal to $j\omega L$. Therefore $\omega L = 10^6$, so $L = 1$ H. At resonance, the impedances of the inductor and capacitor are equal in magnitude, but opposite in sign, so $\omega L = 1/\omega C$ when $\omega = 10^3$. At this frequency $\omega L = 10^3$ Ω, so $10^3 = 1/\omega C$, giving $C = 1/10^6 = 1$ μF.

Box 5

The impedance has the same real part at the two frequencies, indicating the series connection of a resistance of 100 Ω and a reactive component. The imaginary part of Z is directly proportional to frequency, indicating an inductor. The reactance of the inductor (ωL) is 1000 Ω for $\omega = 10^3$, so $L = 1$ H. As a check we would expect the imaginary part of Z to be 10^6 Ω for $\omega = 10^6$, which is what has been measured.

Answer 11.1

Denote the resistance by R and the inductance by L. By voltage divider action we can write

$$\frac{\mathbf{V}_2}{\mathbf{V}_1} = \frac{R}{(R + j\omega L)} = \frac{1}{(1 + j\omega L/R)}$$

The low-frequency asymptote is

$$|\mathbf{V}_2/\mathbf{V}_1| = 1$$

The high-frequency asymptote is

$$|\mathbf{V}_2/\mathbf{V}_1| = R/\omega L$$

The required plot is shown in Figure A11.1. The intersection of the asymptotes is given by $1 = R/\omega L$, i.e., $\omega = R/L = 10^4/10^{-2} = 10^6$ rad/s.

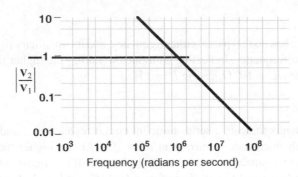

Figure A11.1

Answer 11.3

Figure A11.3 shows the complex quantities involved in our calculations.

Let Z be the impedance of R and C in parallel. Then, from our treatment of the inverter in Chapter 7 (see equation 7.3) we can write

$$\mathbf{V}_{\text{out}} = -(Z/R_A)\mathbf{V}_{\text{in}}$$

Now $Z = R(1/j\omega C)/(R + 1/j\omega C)$ so $\mathbf{V}_{\text{out}}/\mathbf{V}_{\text{in}} = -(R/R_A)/(1 + j\omega CR)$.

At low frequencies $|\mathbf{V}_{\text{out}}/\mathbf{V}_{\text{in}}|$ asymptotes to a constant value of R/R_A which must be chosen to have a value of 15.

At high frequencies $|\mathbf{V}_{\text{out}}/\mathbf{V}_{\text{in}}|$ asymptotes to the function $R/\omega CRR_A = 1/\omega CR_A$.

The two asymptotes intersect at $\omega = 1/CR$ so that C and R must be chosen to satisfy the requirement

$$1/CR = 2\pi \times 100 \times 10^3.$$

Many designs are possible: one is $C = 1$ nF, $R = 1.59$ kΩ, $R_A = 106$ ohms. The corresponding gain asymptotes are seen in Figure A11.3.

Gain at low frequencies is controlled by the ratio R/R_A. The cut-off frequency is determined by R and C.

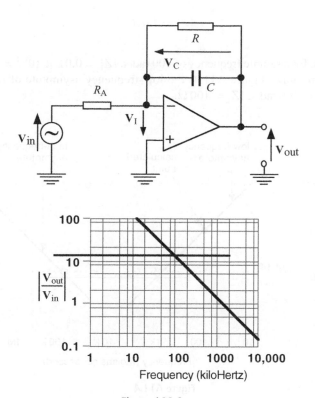

Figure A11.3

Answer 11.4

The impedance Z of the series connection of R, L and C can be expressed as

$$Z = R + j\omega L + 1/j\omega C$$

For high frequencies, where the inductor's contribution to the impedance is dominant, the asymptote of $|Z|$ is given by

$$|Z| = \omega L$$

which gives, for a sample frequency of 10^7 rad/s, $|Z| = 10^7 \times 10^{-4} = 1000\,\Omega$. The high frequency asymptote of $|Z|$ is drawn in Figure A11.4 at 45° (since horizontal and vertical scales are the same) through the point $\omega = 10^7$ rad/s, $|Z| = 1000\,\Omega$.

For low frequencies, where the capacitor's contribution is dominant, the asymptote of $|Z|$ is given by

$$|Z| = 1/\omega C$$

which gives, for a sample frequency of 100 rad/s, $|Z| = 0.01 \times 10^{-4} = 100\,\Omega$. We can therefore draw (Figure A11.4) the low-frequency asymptote of $|Z|$ through the point $\omega = 100\,\text{rad/s}$, $|Z| = 100\,\Omega$.

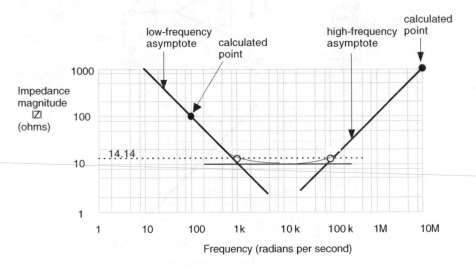

Figure A11.4

If there are frequencies for which the resistance is dominant, then $|Z| = R$ at those frequencies. This relation describes the asymptote $|Z| = 10\,\Omega$ shown in Figure A11.4 from which it can be seen that R is indeed the dominant contribution to $|Z|$ over a range of frequencies for which $R \gg \omega L$ and $R \gg 1/\omega C$.

At the intersection of the mid- and high-frequency asymptotes the effect of the capacitor is so small it can be neglected. Thus, to a very good approximation, $Z = R + j\omega L$. At the intersection $|Z| = R$ and $|Z| = \omega L$, so $\omega = R/L = 10^5$ rad/s. At this frequency $Z = R - jR = R(1 - j)$ and $|Z| = R\sqrt{2} = 14.14\,\Omega$.

Similarly, at the intersection of the mid- and low-frequency asymptotes the effect of the inductor is so small that it can be neglected. Thus, to a very good approximation, $Z = R + 1/j\omega C$. At the intersection $|Z| = R$ and $|Z| = 1/\omega C$ so $\omega = 1/CR = 1000$ rad/s. At this frequency $Z = R - jR = R(1 - j)$ and $|Z| = R\sqrt{2} = 14.14\,\Omega$.

These two points are plotted in Figure A11.4. Together with the three asymptotes they allow one to sketch the variation of $|Z|$ with frequency with an accuracy that is sufficient for initial design purposes.

Answer 11.6

With negative feedback from the opamp output to its negative input terminal we assume a virtual short circuit between the opamp's input terminals. We can therefore write that

$$I_{IN} = V_{IN}/(R_1 + 1/j\omega C_1)$$

The admittance of R_2 and C_2 in parallel is $Y = 1/R_2 + j\omega C_2$ so that the impedance is

$$Z = R_2/(1 + j\omega C_2 R_2)$$

The voltage V_O is therefore equal to $-I_{IN} Z$ which can be expressed as

$V_O/V_{IN} = -[1/(R_1 + 1/j\omega C_1)] \times [R_2/(1 + j\omega C_2 R_2)]$ which simplifies to

$$\frac{V_O}{V_{IN}} = \left(-\frac{R_2}{R_1}\right) \frac{1}{1 + 1/j\omega C_1 R_1} \cdot \frac{1}{1 + j\omega C_2 R_2}$$

At mid-frequencies $|V_O/V_{IN}|$ asymptotes to R_2/R_1 (recall the gain of an inverter).

At low frequencies where the term involving C_1 is dominant, the corresponding asymptote is

$$|V_O/V_{IN}| = \omega C_1 R_2$$

and intersects the mid-frequency asymptote at $\omega - 1/C_1 R_1$.

At high frequencies where the term involving C_2 is dominant, the corresponding asymptote is

$$|V_O/V_{IN}| = 1/\omega C_2 R_1$$

and intersects the mid-frequency asymptote at $\omega = 1/C_2 R_2$

At this point it is useful to lay three straight edges on the specification, two of them at 45°, and move them around until the actual variation of $|V_O/V_{IN}|$ with frequency, as suggested by the asymptotes, appears to satisfy the specification.

Two of many possible designs are shown in Figure A11.6. Design A has a mid-frequency gain of 10, and intersections at 10 and 100 kHz. For this design,

$$2\pi \times 10^4 = 1/C_1 R_1$$

and $2\pi \times 10^5 = 1/C_2 R_2$

Many choices of C_1 and R_1 are possible: we choose $R_1 = 2\,\mathrm{k\Omega}$ and $C_1 = 7.96\,\mathrm{nF}$.

Since R_2/R_1 has been chosen to be equal to 10, $R_2 = 20\,\mathrm{k\Omega}$. The corresponding value of C_2 is therefore $79.6\,\mathrm{nF}$.

In Figure A11.6 an alternative design (B) is illustrated for which the mid-frequency gain is 5 and the asymptote intersections occur for frequencies different from those associated with design A. To confirm the validity of the designs it would be normal to calculate the actual value of $|V_O/V_{IN}|$ at the 'intersection frequencies' to make sure that they lie within the acceptable (shaded) region.

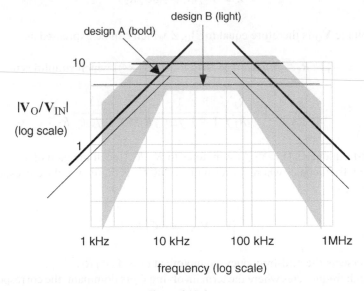

Figure A11.6

Answer 12.1

When $V = 10\,\mathrm{V}$, the current in the $1\,\mathrm{k\Omega}$ resistor is $(20 - 10)/1 = 10\,\mathrm{mA}$

Allow a minimum current of 1 mA through the Zener diode. Then, the maximum current in R is $I = 10 - 1 = 9\,\mathrm{mA}$, corresponding to a value of R equal to $(10\,\mathrm{V})/9\,\mathrm{mA} = 1.11\,\mathrm{k\Omega}$.

If $V = 10$ V and $R = 5$ kΩ then, by Ohm's law, $I = 2$ mA so that the current in the Zener is, by KCL, $10 - 2 = 8$ mA, sufficient to maintain V at 10 V.

If $V = 10$ V and $R = 2$ kΩ, $I = 5$ mA so that the diode current is $10 - 5 = 5$ mA, sufficient to maintain V at 10 V.

If $V = 10$ V and $R = 0.5$ kΩ, I *would* be 20 mA, but this exceeds the 10 mA available through the 1 kΩ resistor (if $V = 10$). Instead, the value of V is found by assuming zero current through the Zener diode and applying the voltage divider principle to calculate V to be $[0.5/(0.5 + 1)] \, 20 = 6.66$ V. As a check on our assumption, the current through the Zener is zero at this value of V.

The change equivalent circuit is shown in Figure A12.1. Neglecting the 20 kΩ resistor in parallel with 10 Ω we see, by using the voltage divider principle, that, to a good approximation, $\Delta V = (10/(1010)$ which is approximately 20 mV.

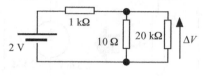

Figure A12.1

Answer 12.2

For $I = 10$ mA, and with 1 mA minimum flowing in the Zener diode, the current through R is 11 mA. Hence, by Ohm's law, $R = (10 - 6)/11 = 364 \, \Omega$.

When $R_L = 3$ kΩ and $V = 6$ V, $I = 2$ mA and the total current through the Zener is $11 - 2 = 9$ mA.

When $R_L = 400 \, \Omega$ then, if V were to be 6 V, I would be equal to 15 mA which is more than would be available through R if V were maintained at 6 V. The only possible state of the circuit is to assume that the Zener current is zero and therefore V can be calculated by the voltage divider principle as $400/(364 + 400) = 5.26$ V.

When the 10 V source increases by 2 V the change circuit is as shown in Figure A12.2(a). To a good approximation, $\Delta V = [10/(10 + 364)] \times 2 = 53$ mV.

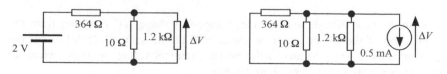

Figure A12.2

When the current drawn by the load increases by 0.5 mA the change circuit is as shown in Figure A12.2(b), from which a good approximation to ΔV can be calculated by the voltage developed across a 10 Ω resistor by a current of

0.5 mA, i.e., −5 mV. The minus sign indicates that V decreases as the load current increases.

Answer 12.3

By Ohm's law the current through the 5 kΩ resistors is $(20 − 10)/(5 + 5) = 1$ mA. The voltage V is therefore $20 − (5\ \text{k}\Omega \times 1\ \text{mA}) = 15$ V.

When the current source is connected to the circuit the change circuit is as shown in Figure A12.4(a). From the circuit we calculate, by Ohm's law, that $\Delta V = 1.25$ V.

As a check, the new currents and voltages are shown in Figure A12.4(b).

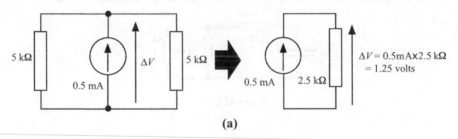

(a)

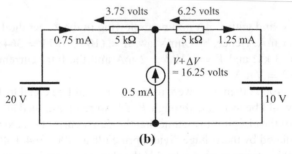

(b)

Figure A12.4

With the new voltage sources, the change in voltage ΔV resulting from the connection of the 0.5 mA current source will be the same (1.25 V) as for the original circuit, simply because in the change circuit the voltage sources – whatever their values – are replaced by short-circuits.

Answer 13.2

Assume the voltage across each diode is 0.7 V.

By Ohm's law, current in left-hand diode $= (4 − 0.7)/330 = 10$ mA: assumption justified.

By Ohm's law, current in right-hand diode $= (4 - 0.7)/3.3$ k$\Omega = 1$ mA: assumption justified.

For the left-hand diode $r_d = 25$ mV/10 mA $= 2.5$ Ω.

For the right-hand diode $r_d = 25$ mV/1 mA $= 25$ Ω.

The small-signal equivalent circuit is shown in Figure A13.2.

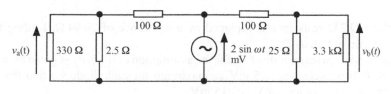

Figure A13.2

Because 330 $\Omega \gg 2.5$ Ω and 3.3 k$\Omega \gg 25$ Ω we can write, to a good approximation, that, by the voltage divider principle, $v_a(t) = (2.5/102.5) \times 2 \sin \omega t$ mV $= 0.049 \sin \omega t$ mV. Similarly, $v_b(t) = (25/125) \times 2 \sin \omega t$ mV $= 0.4 \sin \omega t$ mV.

The assumption of linearity is valid if the amplitude of $v_a(t)$ and $v_b(t)$ is much less than 25 mV. Assume a 'safe' limit of 5 mV.

The largest voltage amplitude is $v_b(t)$, equal to 0.4 mV. Thus, an amplitude of 5 mV at $v_b(t)$ would, in view of the linearity of the small-signal circuit, require a source amplitude of $(5/0.4) \times 2 = 25$ mV. With this source amplitude the amplitude of $v_a(t)$ would be much less than 25 mV. Thus, the maximum source amplitude for the linearity assumption to be valid for both diodes is 25 mV.

Answer 13.3

Consider $V = 5$ V.

Assume that there is sufficient current in the diodes for there to be 0.7 V across each one. Then, by Ohm's law, I_D for each diode is $(5 - 2.1)/5 = 0.58$ mA, thereby justifying the assumption.

Incremental resistance is therefore $r_d = 25$ mV/0.58 mA $= 43.1$ Ω.

The small-signal equivalent circuit is shown in Figure A13.3. It is important to note that the voltage source V is replaced by a short-circuit in the small-signal equivalent circuit. The parallel connection of 129.3 Ω and 5 kΩ is 126 Ω. Therefore, by applying the voltage divider principle the voltage gain is given by

$$v_{out}/v_{in} = 126/1126 = 0.111$$

Now consider $V = 20$ V.

Using the same approach as for $V = 5$ V we find $I_D = 3.58$ mA and $r_d = 6.98$ Ω. The small-signal equivalent circuit has the same form as in Figure A13.3,

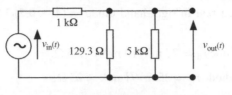

Figure A13.3

but the 129.3 Ω resistance is replaced by a resistance of 20.94 Ω, leading to a voltage gain of 0.0204.

The maximum magnitude of v_O for the assumption of linearity is given by $v_d/25$ mV $\ll 1$. If we select $v_d = 5$ mV as maximum for each diode we find that the maximum v_O is 3 times 5 mV, i.e., 15 mV.

Answer 13.4

Assume 0.7 V across each diode. Then, applying KCL at X we obtain

$$\frac{(12 - V_X)}{2} = \frac{(V_X - 2.8)}{1} + \frac{V_X}{2}$$

giving $V_X = 4.4$ V

Thus, current through each diode is $(V_X - 2.8)/1 = 1.6$ mA, justifying the assumption about diode voltages.

The small-signal model of each diode is a linear resistor given by $r_d = 25$ mV/1.6 mA $= 15.625$ Ω. The small-signal equivalent circuit and its simplified form is therefore as shown in Figure A13.4.

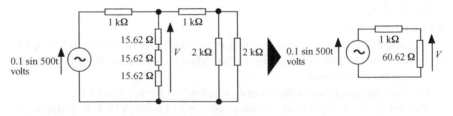

Figure A13.4

By voltage divider action

$$V = (60.62/1060.62) \times 0.1 \sin 500\,t \text{ V} = 0.0057 \sin 500\,t \text{ V}.$$

So the peak-to-peak amplitude of V is 11.4 mV.

Index

Printed and bound by CPI Group (UK) Ltd, Croydon, CR0 4YY